Chemokines Updates

Edited by Murat Şentürk

Published in London, United Kingdom

Chemokines Updates
http://dx.doi.org/10.5772/intechopen.100950
Edited by Murat Şentürk

Contributors
Mohammed El-Gedamy, Vivekanandhan Aravindhan, Srinivasan Yuvaraj, Raghdah Maytham Hameed, Haidar Abdul Amir Najim Abood, Mohanad Mohsin Ahmed, Marija Petrushevska, Dragica Zendelovska, Emilija Atanasovska, Katerina Spasovska, Aleksandar Eftimov, Ion Rogoveanu, Petra Surlin, Dorin Nicolae Gheorghe, Dora Maria Popescu, Flavia Nicolae, Allma Pitru, Alexandra Livia Roman, Vasilica Toma, Liliana Georgeta Foia, Cristina Florescu, Andrada Soanca, Lucian Paul Dragomir, Asuman Akkaya Fırat, Figen Öngöz Dede, Şeyma Bozkurt Doğan

First published in London, United Kingdom, 2023 by IntechOpen
IntechOpen is the global imprint of INTECHOPEN LIMITED, registered in England and Wales, registration number: 11086078, 5 Princes Gate Court, London, SW7 2QJ, United Kingdom
Printed in Croatia

British Library Cataloguing-in-Publication Data
A catalogue record for this book is available from the British Library

Additional hard and PDF copies can be obtained from orders@intechopen.com

Chemokines Updates
Edited by Murat Şentürk
p. cm.

This title is part of the Biochemistry Book Series, Volume 40
Topic: Proteomics
Series Editor: Miroslav Blumenberg
Topic Editors: Paolo Iadarola and Simona Viglio

Print ISBN 978-1-80356-173-8
Online ISBN 978-1-80356-174-5
eBook (PDF) ISBN 978-1-80356-175-2
ISSN 2632-0983

IntechOpen Book Series

Biochemistry

Volume 40

Aims and Scope of the Series

Biochemistry, the study of chemical transformations occurring within living organisms, impacts all of the life sciences, from molecular crystallography and genetics, to ecology, medicine and population biology. Biochemistry studies macromolecules - proteins, nucleic acids, carbohydrates and lipids –their building blocks, structures, functions and interactions. Much of biochemistry is devoted to enzymes, proteins that catalyze chemical reactions, enzyme structures, mechanisms of action and their roles within cells. Biochemistry also studies small signaling molecules, coenzymes, inhibitors, vitamins and hormones, which play roles in the life process. Biochemical experimentation, besides coopting the methods of classical chemistry, e.g., chromatography, adopted new techniques, e.g., X-ray diffraction, electron microscopy, NMR, radioisotopes, and developed sophisticated microbial genetic tools, e.g., auxotroph mutants and their revertants, fermentation, etc. More recently, biochemistry embraced the 'big data' omics systems. Initial biochemical studies have been exclusively analytic: dissecting, purifying and examining individual components of a biological system; in exemplary words of Efraim Racker, (1913 –1991) "Don't waste clean thinking on dirty enzymes." Today, however, biochemistry is becoming more agglomerative and comprehensive, setting out to integrate and describe fully a particular biological system. The 'big data' metabolomics can define the complement of small molecules, e.g., in a soil or biofilm sample; proteomics can distinguish all the proteins comprising e.g., serum; metagenomics can identify all the genes in a complex environment e.g., the bovine rumen.

This Biochemistry Series will address both the current research on biomolecules, and the emerging trends with great promise.

Meet the Series Editor

Miroslav Blumenberg, Ph.D., was born in Subotica and received his BSc in Belgrade, Yugoslavia. He completed his Ph.D. at MIT in Organic Chemistry; he followed up his Ph.D. with two postdoctoral study periods at Stanford University. Since 1983, he has been a faculty member of the RO Perelman Department of Dermatology, NYU School of Medicine, where he is codirector of a training grant in cutaneous biology. Dr. Blumenberg's research is focused on the epidermis, expression of keratin genes, transcription profiling, keratinocyte differentiation, inflammatory diseases and cancers, and most recently the effects of the microbiome on the skin. He has published more than 100 peer-reviewed research articles and graduated numerous Ph.D. and postdoctoral students.

Meet the Volume Editor

Dr. Murat Şentürk received his BS in chemistry in 2002, his MA in biochemistry in 2006, and his Ph.D. in biochemistry in 2009 from Atatürk University. Dr. Şentürk is currently a professor in the Department of Biochemistry in the Faculty of Pharmacy at Ağrı İbrahim Çeçen University. Dr. Şentürk has published over 130 scientific articles, reviews, and book chapters and has presented at various conferences to scientists. His research interests include enzyme inhibitors or activators, protein expression, protein purification and characterization, cancer biochemistry, drug design and synthesis, toxicology, and pharmacology. His research studies have focused on the antioxidant and metabolic enzyme systems. Dr. Şentürk serves on the editorial boards of many international journals.

Contents

Preface

Chemokines (CKs) are a large group of chemotactic cytokines that control cell location, migration, and interactions between different tissues. They are mostly known for their ability to stimulate the migration of cells, especially leukocytes. CKs play a central role in the development and homeostasis of the immune system. They also have essential roles in all destructive or protective immune and inflammatory responses.

CKs play crucial roles in various viral infections, such as influenza, hepatitis B virus (HBV), hepatitis C virus (HCV), viral meningitis, human immunodeficiency virus (HIV), and SARS-CoV-2. CKs mediate the directing of the transport of leukocyte cells into the tumor microenvironment to generate the host response against cancer. CKs can directly modulate tumor tissue expansion by inducing the proliferation of cancerous cells and inhibiting their apoptosis. They can also indirectly modulate the growth of tumor tissue through the effects of CKs on tumor stromal cells by inducing the release of growth and angiogenic factors of cells that make up the tumor microenvironment.

Chemokines Updates contains chapters focusing on research in medical biology and genetics, molecular aspects of medical biotechnology, cancer biochemistry, periodontal diseases, and medical applications in related approaches. The book reviews critical issues and some of the latest developments in applied medical research. Special attention is paid to the clinical aspects of chemokines in the applied health sciences. The book's primary audience is students, researchers, dentists, pharmacists, medical practitioners, and professionals interested in similar fields.

The book has been written by international scientists with expertise in the natural and health sciences, and I would like to thank all the authors for their contributions. We hope the book will increase scientists' knowledge of the complexity of some medical approaches and will encourage both professionals and students to devote some of their future research to understanding the mechanisms and practices involved.

Dr. Murat Şentürk
Professor,
Faculty of Pharmacy,
Ağri Ibrahim Çeçen University,
Ağri, Türkiye

Chapter 1

Chemokines Effective on Platelet Functions

Asuman Akkaya Fırat

Abstract

Chemokines or chemotactic cytokines are chemical signaling molecules that have a regulatory effect on the orientation of endothelial and epithelial cells, especially leukocytes, immune and inflammatory response, and cell regeneration. They are important in the management of endothelial damage, physical harm, atherosclerosis, vascular injury, bleeding, coagulation, interneuron transmission, and platelet functions. Chemokines are divided into four main subfamilies: CXC, CC, CX3C, and C. All of these proteins exert their biological effects by interacting with G-protein-coupled transmembrane receptors called chemokine receptors, which are selectively present on the surfaces of their target cells. Platelet chemokines increase the recruitment of various hematopoietic cells to the vascular wall by nurturing processes, such as neointima formation, atherosclerosis, and thrombosis, while also promoting vessel repair and regeneration after vascular injury. Regarding platelets, CXCL4 (platelet factor 4 and PF4) and the chemokine CXCL7, which is processed from platelet basic protein to connective tissue activating peptide-III and β-thrombomodulin, to its active form neutrophil-activating peptide-2, which are the most abundant. In this chapter, chemokines that are more effective on platelets will be discussed.

Keywords: chemokines, platelets, platelet-activating factor (PAF)

1. Introduction

Chemokines or chemotactic cytokines are chemical signaling molecules that have a regulatory effect on the orientation of endothelial and epithelial cells, especially leukocytes, immune and inflammatory response, and cell regeneration. They are important in the management of endothelial damage, physical damage, atherosclerosis, vascular injury, bleeding, coagulation, interneuron transmission, and platelet functions. Chemokines are divided into four main subfamilies: CXC, CC, CX3C, and C. All of these proteins exert their biological effects by interacting with G-protein-coupled transmembrane receptors called chemokine receptors, which are selectively present on the surfaces of their target cells. Platelet chemokines increase the recruitment of various hematopoietic cells to the vascular wall by nurturing processes, such as neointima formation, atherosclerosis, and thrombosis, while also promoting vessel repair and regeneration after vascular injury. Regarding platelets, CXCL4 (platelet factor 4 and PF4) and the chemokine CXCL7, which is processed from platelet basic protein to connective tissue activating peptide-III and β-thrombomodulin to its active form neutrophil-activating peptide-2, which are the most abundant [1–5]. In this chapter, chemokines that are more effective on platelets will be discussed.

2. Chemokines

The word "chemokine" comes from the ancient Greek word "alchemy" and "kinesis." Cytokines are a family of small cytokines or signaling proteins that are secreted by different cells, especially immune system cells, and induce the movement, communication, and secretory functions of other cell types, including leukocytes as well as endothelial and epithelial cells [1, 2]. In addition to playing an important role in the activation of host immune responses, chemokines are important in biological processes, such as morphogenesis, hemostasis, wound healing, and also in the pathogenesis of diseases, such as cancer [1–3]. Chemokines are classified according to their behavioral and structural properties. All chemokines are small molecules, about 8–10 kDa by mass. The amino acid number and sequence of different chemokine molecules are 20–50% the same. It has four cysteine residues that are the basis for creating the Greek-key-like 3D shapes in basal positions. The first two cysteines in a chemokine are located near the N-terminal end of the protein, the third cysteines in the center of the molecule, and the fourth is near the C-terminal region. This is followed by a single-turn helix called a 3_{10}-helix, three β-strands, and a C-terminal α-helix. These helices and strands are connected by turns called the 30s, 40s, and 50s loops; the third and fourth cysteines are located in the 30s and 50s loops [4]. Chemokines are found in all vertebrates, some viruses, and some bacteria, but none have been found in other invertebrates [5].

Members of the chemokine family are divided into four groups based on the framework formed by the first two cysteine residues:

1. C chemokines

2. CXC chemokines

3. CC chemokines

4. CX_3C chemokines [4]

C chemokines (or γ-chemokines) differ from all other chemokines in that it contains only two cysteines; an N-terminal cysteine and a second cysteine downstream. Two chemokines have been identified for this subgroup and are designated XCL1 (lymphotactin-α) and XCL2 (lymphotactin-β). Thus, the terminology of chemokines is, for example, CCL1 for ligand 1 of the CC-family of chemokines and CCR1 for its corresponding receptor [4].

CXC chemokines the two N-terminal cysteines (or α-chemokines) are separated by an amino acid represented by an "X." There are 17 different CXC chemokines identified in mammals, with a specific amino acid sequence *glutamic acid-leucine-arginine (ELR domain)* just before the first cysteine of CXC. Chemokines with an ELR extension are called ELR-positive and those without an ELR extension are called ELR-negative. ELR-positive chemokines, specifically induce the migration of neutrophils. An example of an ELR-positive CXC chemokine is interleukin-8 (IL-8), which induces neutrophils to leave the bloodstream and enter surrounding tissue. Another example is CXCL13, which is ELR negative and tends to be chemotactic for lymphocytes. Seven of the CXC chemokines have been discovered to date, and these ligands interact with their receptors, CXCRs. Receptors called CXCR1-7 belong to this group of chemokines [6].

CC chemokines (or β-chemokine) proteins have two adjacent cysteines (amino acids), near their amino terminus. There have been at least 27 distinct members of this subgroup reported for mammals, called CC chemokine ligands (CCL)-1 to −28; CCL10 is the same as CCL9. Chemokines of this subfamily usually contain four cysteines (C4-CC chemokines), but a small number of CC chemokines possess six cysteines (C6-CC chemokines). C6-CC chemokines include CCL1, CCL15, CCL21, CCL23, and CCL28 [4–7]. CC chemokines induce the migration of monocytes and other cell types, such as NK cells and dendritic cells. Examples of CC chemokine include monocyte chemoattractant protein-1 (MCP-1 or CCL2), which induces monocytes to leave the bloodstream and enter the surrounding tissue to become tissue macrophages. CCL5 (or RANTES) attracts cells, such as T cells, eosinophils, and basophils, that express the receptor CCR5. Increased CCL11 levels in blood plasma are associated with aging (and reduced neurogenesis) in mice and humans [7].

As a result, cell movement is achieved [1]. Chemokines, according to their amino acid composition, especially are grouped according to the first two cysteine residues of a conserved tetra-cysteine motif: the **CC** and **CXC** form of chemokines are the two largest groups. For example, CX3CL1, XCL1, and XCL2 are named. There are 47 known chemokines, and 19 chemokine receptors [8]. Chemokines that increase leukocyte migration include CCL14, CCL19, CCL20, CCL21, CCL25, CCL27, CXCL12, and CXCL13. Inflammatory provocateurs (such as IL-1 and TNF-alpha) include CXCL-8, CCL2, CCL3, CCL4, CCL5, CCL11, and CXCL10 [6]. This classification is not rigid; for example, CCL20 may also act as a proinflammatory chemokine [5].

3. Receptors

For the cell to respond to a chemokine, it must have a specific chemokine receptor "R." Following binding to the chemokine receptor, it associates with G proteins to transmit cell signals. Activation of G proteins by chemokine receptors then causes activation of an enzyme known as phospholipase C (PLC). PLC cleaves a molecule called phosphatidyl inositol (4,5)-biphosphate (PIP2) into two-second messenger molecules known as inositol triphosphate (IP3) and diacylglycerol (DAG), which trigger intracellular signaling; DAG activates another enzyme called protein kinase C (PKC) and IP3 initiates calcium release from intracellular stores. These reactions trigger multiple signaling cascades (such as the MAP kinase pathway) that produce responses, such as chemotaxis, degranulation, their lease of superoxide anions, and changes in the affinity of cell adhesion molecules called integrins, within the cell that host the chemokine receptor. Chemokine receptors usually belong to the broad group of receptors attached to G-protein (GPCRs). Related to chemokine a calcium signaling cascade is created by binding to its receptor and then causes the activation of small GTPases. This then has downstream activation of integrins (adhesion molecules in the cell) affects and promotes actin polymerization. A pseudopod (cellular projection) appears with polarized cell morphology [5–7].

Homeostatic chemokines are produced continuously in the thymus and lymphoid tissues. It is the chemokines CCL19 and CCL21 (expressed in lymph nodes and on lymphatic endothelial cells) that undertake homeostatic functions in homing, and both have the same receptor CCR7. As a result of binding with these ligands, it is possible to direct antigen-presenting cells (APC) to lymph nodes during the adaptive immune response. Other homeostatic chemokine receptors include CCR9, CCR10, and CXCR5. CCR9 promotes the migration of leukocytes to the gut, CCR10 to the skin, and CXCR5 provokes the migration of B cells to the follicles of the lymph nodes. In addition, CXCL12 (SDF-1: stromal cell-derived factor 1), which is produced structurally in the bone marrow, supports the proliferation of progenitor B cells in the bone marrow microenvironment [7, 8]. The protein encoded by this gene is a member of the CXC chemokine family and is a majör mediator of the inflammatory response. The encoded protein is commonly referred to as interleukin-8 (C-X-C motif chemokine ligand 8). IL-8 is secreted by mononuclear macrophages, neutrophils, eosinophils, T lymphocytes, epithelial cells, and fibroblasts. Bacterial and viral products induce IL-8 expression. IL-8 also participates with other cytokines in the pro-inflammatory signaling cascade and plays a role in systemic inflammatory response syndrome (SIRS). This gene is believed to play a role in the pathogenesis of the lower respiratory tract infection bronchiolitis, a common respiratory tract disease caused by their respiratory syncytial virus (RSV) [8–10]. The overproduction of this pro-inflammatory protein is thought to cause lung inflammation associated with cystic fibrosis. This pro-inflammatory protein is also suspected of playing a role in coronary artery disease and endothelial dysfunction. This protein is also secreted by tumor cells and promotes tumor migration, invasion, angiogenesis, and metastasis. This chemokine is also a potent angiogenic factor [8–10]. The binding of IL-8 to one of its receptors (IL-8RB/CXCR2) increases the permeability of blood vessels and increasing levels of IL-8 are positively correlated with increase severity of multiple disease outcomes (e.g., sepsis). This gene and other members of the CXC chemokine gene family form a gene cluster in a region of chromosome 4q [7–11]. Chemokines, according to the types of cells affect monocytes/macrophages. Chemokines that mobilize these cells toward the site of inflammation include CCL2, CCL3, CCL5, CCL7, CCL8, CCL13, CCL17, and CCL22 [9, 12].

3.1 T-lymphocytes

Chemokines that attract T lymphocytes to the site of inflammation are: CCL2, CCL1, CCL22, and CCL17. In addition, CXCR3 expression is mediated by activated T cells. IFN-y-inducible chemokines are CXCL9, CXCL10, and CXCL11 [9, 12].

3.2 Mast cells

Express receptors for chemokines on their surface are: CCR1, CCR2, CCR3, CCR4, CCR5, CXCR2, and CXCR4. The ligands of these receptors CCL2 and CCL5 play an important role in mast cell recruitment and activation in the lung. There is also evidence that CXCL8 can inhibit mast cells [9–12].

3.3 Eosinophils

The migration of eosinophils various tissues is provoked by several chemokines of the CC family, which are as follows: CCL11, CCL24, CCL26, CCL5, CCL7, CCL13, and CCL3. The chemokines CCL11 (eotaxin) and CCL5 (RANTES) act via a specific CCR3 receptor on the surface of eosinophils, and eotaxin plays an important role in the initial recruitment of eosinophils to the lesion [9–12].

3.4 Neutrophils

Regulated primarily by CXC chemokines. CXCL8 (IL-8) is particularly chemotactic for neutrophils and activates their metabolism and degranulation [10–12].

Platelets are nuclear cell fragments derived from megakaryocytes that contain both pro-inflammatory and (anti-inflammatory) fragments in abundance. In particular, they play important roles in hemostasis. Angiogenic mediators physiological and pathological conditions in the vascular system, and immune cells are important in regulating platelet functions. Platelets as the main players in thrombosis and hemostasis. It is becoming a more interesting topic with increasing discoveries in the composition of inflammatory and immune-modulating molecules [10–12]. An important phenomenon also emerges in the atherosclerosis of platelets. It found an association between increased platelet concentration, aggregation, and the long-term incidence of fatal coronary heart disease in a population of apparently healthy middle-aged men [12]. Many platelet-derived substances and chemokines family of interesting proteins stored in it forms α-granules and exhibits numerous biological activities. In addition, chemokines are reexpressed by platelets [12, 13].

4. Platelet chemokines

The first recognized platelet chemokine PF-4 is now known as CXCL4. TGF beta-1 is also considered a chemokine. The NAP-2 fragment, now called CXCL7, is also a platelet chemokine. CCL3 (MIP-1), CCL5 (RANTES), CCL7 (MCP-3), and CXCL1 are also platelet-related chemokines. CCL17 (TARC) has recently been reported in platelets. CCL17 is an autocrine factor that increases platelet activation and its receptor CCR4 is also found in platelets. When platelets in plasma are activated *in vitro*, serum concentrations of these chemokines reach the range of 1–5 mol/L [13–17] (**Table 1**).

CXCL4 was the first chemokine, whose effects on platelets were reported in 1985. Platelet factor 4 (PF4) is a minor cytokine belonging to the CXC family of chemokines, also known as chemokine ligand 4 (CXCL4). This chemokine is released from

Chemokine	Alternative name	Receptor
CXCL1	GRO-α	CXCR2 > CXCR1
CXCL4	PF4	CXCL3B, GAG
CXCL4L1	PF4alt	Unknown
CXCL5	ENA-78	CXCR2
CXCL7	PBP, βTG, CTAPIII, NAP-2	CXCR > CXCRI
CXCL8	IL-8	CXCRI, CXCR2
CXCL12	SDF-I α	CXCR4
CCL2	MCP-I	CCR2
CCL3	MIP-I α	CCRI, CCR2, CCR3
CCL5	RANTES	CCRI, CCR3, CCR4, CCR5
CCL17	TARC	CCR4, CCR8

Table 1.
The alternative names and receptors of chemokines.

the alpha granules of activated platelets during platelet aggregation and provokes blood coagulation by regulating the effects of heparin-like molecules. Due to these roles, it is predicted that they will also play a role in wound healing and vascular repair [5]. It is usually found in a complex with proteoglycan [18].

The human PF4 gene is located in human chromosome 4. Platelet factor 4 is a 70 amino acid-containing protein that is released from the alpha granules of activated platelet sand and binds to heparin with high affinity. Its main physiological role is thought to be the neutralization of heparin-like molecules on the endothelial surface of blood vessels, thereby inhibiting local antithrombin activity and provoking coagulation. As a potent chemotactic factor for neutrophil sand fibroblasts, PF4 probably has a role in inflammation and wound healing [19, 20]. PF4 is also a chemotactic factor for neutrophils, fibroblasts, and monocytes, and interacts with an additional counterpart of the chemokine receptor CXCR3, known as CXCR3-B [20].

PF4 complex is the antigen in heparin-induced thrombocytopenia (HIT), which is an autoimmune reaction specific to anticoagulant heparin administration [21]. PF4 autoantibodies have also been found in patients with thrombosis and similar to HIT, but who have not been given heparin before [22]. Antibodies against PF4 have been blamed in cases of thrombosis and thrombocytopenia after vaccination with the Oxford-Astra Zeneca or Janssen COVID-19 vaccine [23, 24]. This phenomenon was named vaccine-induced immune thrombotic thrombocytopenia (VITT) [25]. A relationship was also found in PF4 expression with long-term COVID symptoms [26]. It increases in patients with systemic sclerosis, who also have interstitial lung disease [27].

Human platelet factor 4 also specifically decomposes the digestive vacuole of the malaria parasite and neutralizes the malaria parasites in erythrocytes [28].

Transforming growth factor-beta 1 (TGF-beta1) is a polypeptide member of the transforming growth factor beta superfamily of cytokines. It is a secreted protein that performs many cellular functions, including the control of cell growth, proliferation, differentiation, and apoptosis. In humans, TGF-β1 is encoded on the TGFB1 gene [28, 29].

5. CXCL7

TGF-β Transforming growth factor-beta 1 (TGF-beta1) is a polypeptide member of the transforming growth factor beta superfamily of cytokines. It is a secreted protein that performs many cellular functions, including the control of cell growth, proliferation, differentiation, and apoptosis. In humans, TGF-β1 is encoded on the TGFB1 gene [28, 29].

A thrombomodulin (C-X-C motif) ligand 7, β-thrombomodulin (β-TG) or beta-thromboglobulin, is a chemokine protein secreted and stored by platelets [29–31]. Along with platelet factor 4 (PF4), β-TG is one of the themes specific platelet-specific proteins β-TG and PF4 are stored in platelet alpha granules and released during platelet activation [29, 32, 33]. In conclusion, it is a useful marker of platelet activation [29, 32]. β-TG also plays an important role in the maturation of megakaryocytes [34].

Among the chemokines stored and secreted by platelets, CXCL7 is the largest representative. β-TG levels are used as an index of platelet activation. It is measured in blood plasma or urine by the ELISA method and usually together with PF4. When platelets are active, CXCL4 is 0.4–1.9 μM in serum, while CXCL7 is 1.6–4.8 μM [35]. It is used as a measure for platelet activation [36].

It consists of proteolytic derivatives of 128 aa-precursor molecules called pre-PBP, the primary CXCL7 translation product. The primary sequence, pre-PBP is a 34 amino acid residue leader sequence [37]. β-TG is a molecule and N-terminal variant of pro CXCL7 containing 81 residue amino acids. Platelet basic protein (PBP; 94 aa), connective tissue activating peptide III (CTAP-III; 85aa), and neutrophil-activating neutrophil-activating (NAP-2, 70 aa), are CXCL7 N-terminal variants that diverge [38]. β-TG levels increase with age and in diabetes mellitus [39, 40]. β -TG levels were found to increase during treatment with synthetic ethinyl estradiol, but not significantly in that treated with the natural estradiol valerate [41–44]. β-TG levels were also found to be slightly increased or unchanged in an uncomplicated pregnancy [45].

Besides both CTAP-III and NAP-2o, the medium-size shortened CXCL7 variants, all its capacity to support various aspects of fibroblast metabolism has been demonstrated. For example, the synthesis of matrix components, such as hyaluronic acid and glycosaminoglycans (GAGs) [46]. Increased GLUT-1 glucose transporter expression and concomitant cellular glucose uptake are enhanced [47]. As a platelet-derived mediator, CXCL7 may also participate in reparative functions following vascular tissue injury. But the CXCL7, its role as a growth factor, is controversial. Thus, it is more likely that fibroblast mitogenic activity may be the task of full-length-PBP [48]. Like CXCL7, platelet factor 4 (PF4, CXCL4) and its closely related chemokine, platelet basic protein (PBP), are important in platelets. The role of PF4 in hemostasis/thrombosis *in vivo* has been demonstrated, PF4 plays a role in pathological thrombotic conditions, such as heparin-induced thrombocytopenia (HIT) and septic shock [48, 49].

The abundance of CXCL7 variants follows a specific sequence of proteolytic vents during platelet production and activation. Thus, the main player on megakaryocytes is PBP. A small amount of CTAP-III also provides stimulation. The proportion of the shorter variant CTAP-III as platelets mature increases, and the proportion of PBP drops to about 25% [49].

Removal of the inhibitory N-terminus to activate CXCL7 as a neutrophil-directed chemokine underlies its potential role for intravascular and extravascular. As we have shown, the translation of CXCL7 to NAP-2 is mainly catalyzed by NAP-2 target neutrophils. Neutrophil activation occurs via the serine protease cathepsin G-linked plasma inhibitors, which are not effective [50].

6. CXCL1

CXCL1 is a small peptide belonging to the CXC chemokine family that acts as a chemoattractant for several immune cells, especially neutrophils [51] or other non-hematopoietic cells, to the site of injury or infection and plays an important role in the regulation of immune and inflammatory responses. It was previously called the GRO1 oncogene, GROα, neutrophil-activating protein 3 (NAP-3), melanoma growth stimulating activity, alpha (MGSA-α). It is also known as keratinocytes-derived chemokine (KC) in mice or cytokine-induced neutrophil chemoattractant type-1 (CINC-1) in rats. In humans, this protein is encoded by the gene Cxcl1 [5] and is located on human chromosome 4 among genes for other CXC chemokines [52].

Under normal conditions, CXCL1 is not constitutively expressed. It is produced by activated macrophages, neutrophils, and epithelial cells, or by different immune cells, such as Th17. Moreover, its expression is indirectly provoked by IL-1, TNF-α, or IL-17 released by Th17 cells [11]. It plays a major role in inflammation [53, 54].

CXCL1 has a potentially similar effect as interleukin-8 (IL-8/CXCL8). Binds to the CXCR2, receptor CXCL1 triggers phosphatidyl inositol-4,5-biphosphate 3-kinase-y (PI3Kγ)/Akt, MAP kinases, such as ERK1/ERK2 or phospholipase-β (PLCβ) signaling pathways. CXCL1 increases the expression of inflammatory responses, and thus, contributes to the inflammation process [12]. CXCL1 is also involved in wound healing and oncogenesis processes [55–57].

CXCL1 has been shown to have roles in the development of breast cancer, gastric and colorectal carcinoma, or lung cancer tumors [58, 59]. In addition, it has been reported that CXCL1 is secreted by human melanoma cells and plays a role in mitogenic activity [60–62].

CXCL1 is expressed by neurons and oligodendrocytes in the brain and spinal cord and by microglia during pathologies, such as Alzheimer's disease, multiple sclerosis, and brain damage. A study in mice shows evidence that CXCL1 reduces the severity of multiple sclerosis [23]. In addition, CXCL1 contributes to CXCL1, playing a role in spinal cord development by acting on oligodendrocytes [7]. CXCR2 receptors for s to the release of prostaglandins, thereby resulting in increased sensitivity to pain. It initiates nonspecific sensitivity through the recruitment of neutrophils into the tissue. It increases the transcription of genes that induce chronic pain, such as cyclooxygenase-2 (COX-2) [12].

7. CCL 3

Chemokine (C-C motif) ligand 3 (CCL3) also known as macrophage inflammatory protein 1-alpha (MIP-1-alpha), is located on the CCL3 gene in humans [3]. By binding to all of the CCL3, CCR1, CCR4, and CCR5 receptors, it may play a role in the the recruitment and activation of polymorphonuclear leukocytes [63] in acute inflammatory conditions. Sherry et al. showed two protein subcomponents of MIP-1 called alpha (CCL3) and beta (CCL4) [64, 65]. CCL3 can produce rapid-onset symptoms of monophasic fever that are greater than or equal to fevers produced by recombinant human tumor necrosis factor or recombinant human interleukin-1. Moreover, unlike these two endogenous pyrogens, MIP-1-induced fever is capable of producing cyclooxygenase-induced fever. It is not inhibited by ibuprofen. CCL3 may participate in a type of febrile response that is not produced by prostaglandin and cannot be clinically inhibited by cyclooxygenase. CCL3 has been shown to interact with CCL4 to activate macrophages, monocytes, and neutrophils [66].

CCL4 also known as macrophage inflammatory protein-1β (MIP-1β), is a CC chemokine that specifically binds to CCR5 receptors. It is chemotactic for natural killer cells, monocytes, and various other immune cells [67] CCL4 is an important HIV suppressive factor released by CD8+ T cells [68]. Performance low-memory CD8+ T cells that normally express MIP-1-beta. The concentration of this chemokine is inversely proportional to micro RNA-125b. The concentration of CCL4 in the body increases with age, which can lead to chronic inflammation and liver damage and may be a marker [69].

8. CCL5

Chemokine C-C ligand 5 (CCL5) is a protein encoded on the CCL5 gene in humans [70]. The gene was discovered in situ hybridization in 1990 [71]. Also known as

RANTES (regulated by activation, normal T-cell expressed and secreted). CCL5 belongs to the CC subfamily of chemokines, due to its adjacent cysteines near the N terminus. It is an 8 kDa protein acting as a classical chemotactic cytokine or chemokine. It consists of 68 amino acids. CCL5 is a pro-inflammatory chemokine, that recruit's leukocytes to the site of inflammation. It is chemotactic for T cells, eosinophils, and basophils, but also monocytes, natural killer (NK) cells, dendritic cells, and mastocytes [72]. With the help of particular cytokines (i.e., IL-2 and IFN-γ) that are released by T cells, CCL5 also induces the proliferation and activation of certain NK cells to form CHAK (CC-chemokine-activated killer) cells. It is also an HIV-suppressive factor released from CD8+ T cells [72, 73].

The chemokine CCL5 is mainly expressed by T cells and monocytes, and it is not expressed by B cells. Moreover, it is abundantly expressed by epithelial cells, fibroblasts, and thrombocytes. Although it can bind to receptors CCR1, CCR3, CCR4, and CCR5 belonging to seven transmembrane G-protein-coupled receptor (GPCRs) family [8], it has the highest affinity to the CCR5. CCR5 is presented on the surface of T cells, smooth muscle endothelial cells, epithelial cells, parenchymal cells, and other cell types. After the binding of CCL5 to CCR5, phosphoinositide 3-kinase (PI3K) is phosphorylated and subsequently, the phosphorylated PI3K phosphorylates protein kinase B (PKB; also known as Akt) on the serine 473. Then, the Akt/PKB complex phosphorylates and inactivates a serine/threonine protein kinase GSK-3. After the CCL5/CCR5 binding, some other proteins are regulated as well. Bcl2 is more expressed and it induces apoptosis [74, 75].

RANTES acts as a typical chemokine causing chemotaxis of mononuclear cells at nanomolar concentrations. Transendothelial migration of monocytes and lymphocytes is integrin-dependent and requires adhesion molecules from molecules, such as ICAM-1 and ICAM-1 [76, 77].

Due to resting integrins, tissue needs activation signals confirmation and only has a low affinity for its ligands. Chemokines are important stimuli for integrin activation as they are released during inflammation and induce adhesion in all types of leukocyte subsets. It is phosphorylated shortly after exposure to RANTES [78].

Until now, the mechanism of integrin-dependent adhesion of RANTES has not been fully elucidated. In addition to initiating cell migration at high concentrations, RANTES also acts independently of its G-protein-coupled receptors. It induces the release of T cells and pro-inflammatory mediators [79, 80]. Its unique capacity to form homotypic clusters and its high affinity for GAGs on the surface of endothelial cells, basement membrane and extracellular matrix RANTES will be immobilized [79–82].

AGs heparin-binding site associated with residues located at the C-terminus of PF4 tetramer displays a band of positively charged residues [83]. In the RANTES molecule, two basic amino acid clusters present heparin-binding motifs in the 40s loop between this cond and the third β-strand only in the C-terminal α-helix. The essential remains in the 40s loop are specific and it shows the high affinity of RANTES to different GAG species. Effective leukocyte arrest in the endothelium, especially unlike chemotaxis appears to depend on the formation of RANTES oligomers to bridge the surface-bound RANTES and CCR1 [84]. Selective binding of chemokines to subgroups of glycosaminoglycans cell surface induces polymerization facilitating their attachment it binds to receptor and enhances their effects on high-affinity receptors in the local microenvironment [85, 86]. In addition, structural motifs required for oligomerization RANTES are important for heterophilic interaction. RANTES with PF4 increases surface immobility and enhances monocyte adhesion to endothelial cells under flow conditions [86, 87].

9. Chemokines and vascular biology

Data from animal experiments suggest that activated platelets are involved in the pathogenesis of atherosclerosis, which indicates that it is important in acute thromboembolism [88, 89]. Monocyte recruitment into the subendothelial artery space is an early step in pathogenesis. Platelets with an affinity for GAGs accumulate in the luminal endothelium, where chemokines, such as RANTES activate monocytes [83, 90]. The role of RANTES in humans is reported to be less. In one study group, serum levels of RANTES were found to be lower in atherosclerosis compared to healthy controls 92. PF4 and NAP-2 as well been shown to play a role in atherosclerosis. PF4, which can reflect platelet activity. Studies have been conducted on its importance in atherosclerosis [91–94]. PF4 and, to a lesser extent, NAP-2 are associated with human atherosclerotic plaques [94]. Besides, PF4 is involved in the metabolism of atherogenic lipids, for example, oxidized low-density lipoprotein (LDL). It is stated that LDL (oxLDL) plays a role in the atheroma plaque [94].

Patients with stable and especially unstable angina exhibit markedly elevated plasma NAP-2 levels. PF4 may also play an important role in the acute coronary syndrome that causes plaque formation. But "how does PF4 affect thrombosis?" has not been finally clarified [95]. Angiogenesis, which may be beneficial in wound healing, also it is effective in pathological conditions, such as cancer and atherosclerosis. Capillary sprouting and endothelial cell proliferation VEGF (vascular endothelial growth factor), bFGF (essential fibroblast growth factor) by *in vitro* platelet releases, and PDGF (platelet-derived growth factor) can be induced by isolated platelets [96, 97]. NAP-2 accelerates endothelial cell healing with CXCR2-dependent fibronectin, fibrinogen, and platelet-coated endothelial matrix of endothelial progenitor cells [98]. On the other hand, platelet products PF4 and PF4alt are potent inhibitors of angiogenesis. PF4 can exert its angiostatic activity via CXCR3B [99] (**Figure 1**).

The mechanism mediated by CXCR2 constitutes a very important area of research. CXCR2 binds to a G-protein. Multiple ligands are available. CXCR2 results in activation by binding of chemokines. NFκB, MAPK, PI3K, and Rac 1 are among other

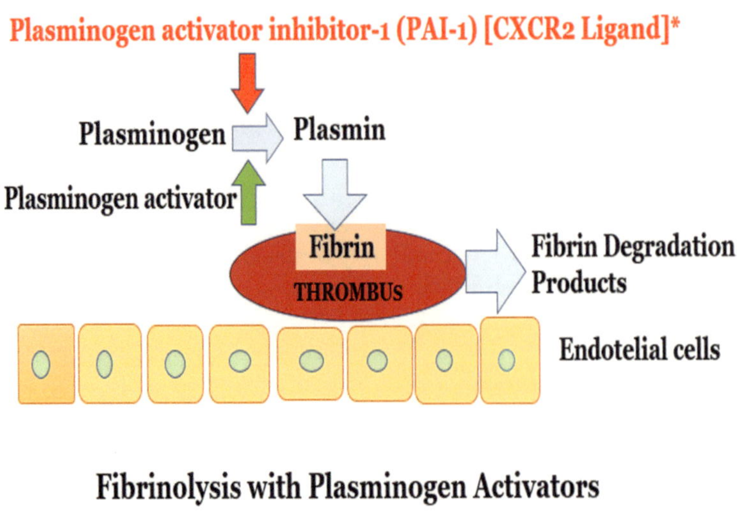

Figure 1.
Fibrinolysis with plasminogen activators.

signaling cascades. Activation by CXCR2 increases NAPDH oxidases, causing an explosion of reactive oxygen species (ROS). This ROS explosion plays a role in clearing pathogen infections by macrophages. It mediates the induction of apoptosis in cancer cells [100–102].

Congenital deficiency of PAI-1; since fibrinolysis is not sufficiently suppressed, it causes hemorrhagic diathesis. PAI-1 is found at increased levels in various disease states (such as several cancer types), as well as obesity and metabolic syndrome. It has been associated with increased thrombosis formation in patients with these conditions. PAI-1 can induce cellular senescence. PAI-1 appears to play an important role in the progression of fibrosis in inflammatory conditions, where fibrin accumulates in tissues. Possibly, lower PAI levels will lead to less suppression of fibrinolysis and conversely faster degradation of fibrin. Angiotensin II increases the synthesis of plasminogen activator inhibitor-1, thereby accelerating the development of atherosclerosis [103, 104].

Thrombotic complications are common in COVID-19 and contribute significantly to mortality and morbidity. Immune-mediated thrombotic mechanisms, complement activation, macrophage activation syndrome, antiphospholipid antibody syndrome, hyperferritinemia, and renin-angiotensin system dysregulation may be potential prognostic biomarkers in COVID-19. Recent studies are currently discussing the hypothetical benefits and anticipated challenges of therapeutic anticoagulation and fibrinolytic therapy in COVID-19 [105].

Abbreviations

CHAK	Chemokine-activated killer
COVID-19	Sars-Cov2 Virus 19
IP3	Inositol triphosphate3
LDL	Low-density lipoprotein
MCP- β	Monocyte chemoattractant protein-1 beta
MGSA	Melanoma growth stimulatingactivity alpha
MIP-1 β	Macrophage inflammatory protein 1beta
NAP3	Neutrophil activating protein-3
NK	Naturel killer
PAI1	Plasminogen activator inhibitor1
PIP2	Phosphatidyl inositol2
PF4	Platelet Factor4
PLC	Phospholipase C
RANTES	Regulated by activation normal Tcell expressed and secreted
ROS	Reactive oxygen species
RSV	Respiratory syncytial virus
SDF1	Stromal cell-derived factor1
SIRS	Systemic inflammatory response syndrome
TNFα	Tumor necrosis factor-alpha

Author details

Asuman Akkaya Fırat
Biochemistry Department, Fatih Sultan Mehmet (F.S.M) Training and Research Hospital, Health Sciences University (S.B.U), İstanbul, Turkey

*Address all correspondence to: asumanfirat44@gmail.com

References

[1] Raman D, Sobolik-Delmaire T, Richmond A. Chemokines in health and disease. Experimental Cell Research. 2011;**317**(5):575-589

[2] Charo IF, Ransohoff RM. The many roles of chemokines and chemokine receptors in inflammation. New England Journal of Medicine. 2006;**354**(6):610-621

[3] Davenport RD. An introduction to chemokines and theirroles in transfusion medicine. Vox Sanguinis. 2009;**96**(3):183-198

[4] Miller MC, Mayo KH. Chemokines form a structural perspective. International Journal of Molecular Sciences. 2017;**18**(10):2088

[5] Fernandez EJ, Lolis E. Structure, function, and inhibition of chemokines. Annual Review of Pharmacology and Toxicology. 2002;**42**:469-499

[6] Matsuo K, Yoshie O, Nakayama T. Multifaceted roles of chemokines and chemokine receptors in tumor immunity. Cancers (Basel). 2021;**13**(23):6132

[7] Villeda SA, Luo J, Mosher KI, Zou B, Britschgi M, Bieri G, et al. The ageing systemic milieu negatively regulates neurogenesis and cognitive function. Nature. 2011;**477**(7362):90-94

[8] Laing KJ, Secombes CJ. Chemokines. Developmental and Comparative Immunology. 2004;**28**(5):443-460

[9] Xie JH, Nomura N, Lu M, Chen SL, Koch GE, Weng Y, et al. Antibody-mediated blockade of the CXCR3 chemokine receptor results in diminished recruitment of T helper 1cells into sites of inflammation. Journal of Leukocyte Biology. 2003;**73**(6):771-780

[10] Reeves EP, Williamson M, O'Neill SJ, Greally P, McElvaney NG. Nebulized hypertonic saline decreases IL-8 in sputum of patients with cystic fibrosis. American Journal of Respiratory and Critical Care Medicine. 2011;**183**(11):1517-1523

[11] Ang Z, Abdi GunawanKoen R, Er JZ, Lee LT, Tam Kit Chung J, Guo H, et al. Novel AU-richproximal UTR sequences (APS) enhance CXCL8 synthesisupon the induction of rpS6 phosphorylation. PLOS Genetics. 2019;**15**(4):e1008077

[12] Ono SJ, Nakamura T, Miyazaki D, Ohbayashi M, Dawson M, Toda M. Chemokines: Roles in leukocyte development, trafficking, and effector function. The Journal of Allergy and Clinical Immunology. 2003;**111**(6):1185-1199

[13] Deuel TF, Keim PS, Farmer M, Heinrikson RL. The amino acid sequence of human platelet factor 4. Proceedings of the National Academy of Sciences of the United States. 1977;**74**:2256-2258

[14] Dobroski DR, Rabbani LE, Loscalzo J. The relationship between thrombosis and atherosclerosis. In: Schafer JLAI, editor. Thrombosis and Hemorrhage. Baltimore, MD: Williams & Wilkins; 1998. pp. 837-861

[15] Dore M. Platelet-leukocyte interactions. The American Heart Journal. 1998;**135**:S146-S151

[16] Brandt E, Ludwig A, Petersen F, Flad HD. Platelet-derived CXC chemokines: Old players in new games. Immunological Reviews. 2000;**177**:204-216

[17] Hundelshausen P, Petersen F, Brandt E. Platelet-derived chemokines in vascular biology. Thrombosis and Hemostasis. 2007;**97**(05):704-713

[18] Eisman R, Surrey S, Ramachandran B, Schwartz E, Poncz M. Structural and functional comparison of the genes for human platelet factor 4 and PF4alt. Blood. 1990;**76**(2):336-344

[19] O'Donovan N, Galvin M, Morgan JG. Physical mapping of the CXC chemokine locus on human chromosome 4. Cytogenetics and Cell Genetics. 1999;**84**(1-2):39-42

[20] Lasagni L, Francalanci M, Annunziato F, Lazzeri E, Giannini S, Cosmi L, et al. An alternatively spliced variant of CXCR3 mediates the inhibition of endothelial cell growth induced by IP-10, Mig, and I-TAC, and acts as a functional receptor for platelet factor 4. The Journal of Experimental Medicine. 2003;**197**(11):1537-1549

[21] Warkentin TE. Drug-induced immune mediated thrombocytopenia—from purpura to thrombosis. The New England Journal of Medicine. 2007;**356**(9):891-893

[22] Warkentin TE, Makris M, Jay RM, Kelton JG. A spontaneous prothrombotic disorders enabling heparin-induced thrombocytopenia. The American Journal of Medicine. 2008;**121**(7):632-636

[23] Schultz NH, Sørvoll IH, Michelsen AE, Munthe LA, Lund-Johansen F, Ahlen MT, et al. Thrombosis and thrombocytopenia after ChAdOx1 nCoV-19 vaccination. The New England Journal of Medicine. 2021;**384**(22):2124-2130

[24] Greinacher A, Thiele T, Warkentin TE, Weisser K, Kyrle PA, EichingerS.Thromboticthrombocytopenia after ChAdOx1 nCov-19 vaccination. The New England Journal of Medicine. 2021;**384**(22):2092-2101

[25] Arepally GM, Ortel TL. Vaccine-induced immune thrombotic thrombocytopenia (VITT): What we know and don't know. Blood. 29 Jul 2021;**138**(4):293-298. DOI: 10.1182/blood.2021012152

[26] Ryan FJ, Hope CM, Masavuli MG, Lynn MA, Mekonnen ZA, Yeow AE, et al. Longterm perturbation of the peripheral immune system months after SARS-CoV-2 infection. BMC Medicine. 2022;**20**(1):26

[27] Volkmann ER, Tashkin DP, Roth MD, Clements PJ, Khanna D, Furst DE, et al. Changes in plasma CXCL4 levels are associated with improvements in lung function in patients receiving immunosuppressive therapy for systemic sclerosis-related interstitial lung disease. Arthritis Research & Therapy. 2016;**18**(1):305

[28] Love MS, Millholland MG, Mishra S, Kulkarni S, Freeman KB, Pan W, et al. Platelet factor 4 activity against P. Falciparum and its translation to nonpeptidic mimics as antimalarials. Cell Host & Microbe. 2012;**12**(6):815-823

[29] Sharma G, Berger JS. Platelet activity and cardiovascular risk healthy individuals: A review of the data. The Journal of Thrombosis and Thrombolysis. 2011;**32**(2):201-208

[30] Kaplan KL, Owen J. Plasma levels of beta-thromboglobulin and platelet factor 4 as indices of platelet activation in vivo. Blood. 1981;**57**(2):199-202

[31] Pillai MM, Iwata M, Awaya N, Graf L, Torok-Storb B. Monocyte-derived CXCL7 peptides in the marrow microenvironment. Blood. 2006;**107**(9):3520-3526

[32] Jumpupto Cella G, Scattolo N, Girolami A, Sasahara AA. Are platelet factor 4 and β-thromboglobulin markers of cardiovascular disorders? Ricerca in Clinic and Laboratory. 1984;**14**(1):9-18

[33] Jumpupto Cytokines and Cells Online Pathfinder Encyclopaedia. Beta-Thromboglobulin Retrieved on August 17, 2009

[34] Ha C-E, Bhagavan NV. Essentials of Medical Biochemistry: With Clinical Cases. Elsevier Inc.: All rights reserved Academic Press; 2015. p. 652 ISBN 978-0-12-095461-2. DOI: 10.1016/C2009-0-00064-6

[35] Loetscher P, Seitz M, Baggiolini M, Moser B. Interleukin-2 regulates CC chemokine receptor expression and chemotactic responsiveness in T lymphocytes. Journal Experimental Medicine. 1996;**184**:569-577

[36] Moore S, Pepper DS, Cash JD. The isolation and characterisation of a platelet-specific β-globulin (β-thromboglobulin) and the detection of antiurokinase and antiplasmin released from thrombin-aggregated washed human platelets. Biochimica et Biophysica Acta. 1975;**379**:360-369

[37] Begg GS, Pepper DS, Chesterman CN, et al. Complete covalent structure of human beta-thromboglobulin. Biochemistry. 1978;**17**:1739-1744

[38] Majumdar S, Gonder D, Koutsis B, et al. Characterization of the human beta-thromboglobulin gene. Comparison with the gene for platelet factor 4. The Journal of Biological Chemistry. 1991;**266**:5785-5789

[39] Martinelli I, Bucciarelli P, Mannucci PM. Thrombotic risk factors: Basic pathophysiology. Critical Care Medicine. 2010;**38**(2 Suppl):S3-S9

[40] Sterne J, Kirkwood BR. Essential Medical Statistics. Oxford: Blackwell Science; 2003

[41] Farris M, Bastianelli C, Rosato E, Brosens I, Benagiano G. Pharmaco dynamics of combined estrogen-progestin oral contraceptives: 2. effects on hemostasis. Expert Review of Clinical Pharmacology. 2017;**10**(10):1129-1144

[42] Kluft C, Lansink M. Effect of oral contraceptives on hemostasis variables. Thrombosis and Hemostasis. 1997;**78**(1):315-326

[43] Kuhl H. Adverse effects of estrogen treatment: Natural vs. synthetic estrogens. In: Lippert TH, Mueck AO, Ginsburg J, editors. Sex Steroid sand the Cardiovascular System: Proceedings of the 1st Interdisciplinary Workshop, Tuebingen, Germany, October 1996. London/New York: Parthenon; 1998. pp. 201-210

[44] Lindberg UB, Crona N, Stigendal L, Teger-Nilsson AC, Silfverstolpe G. A comparison between effects of estradiol valerate and low dose Ethinylestradiol on hemostasis parameters. Thrombosis and Hemostasis. 1989;**61**(1):65-69

[45] Hellgren M. Hemostasis during normal pregnancy and puerperium. Seminars in Thrombosis and Hemostasis. 2003;**29**(2):125-130

[46] Castor CW, Walz DA, Johnson PH, et al. Connective tissue activation. XXXIV: Effects of proteolytic processing on the biologic activities of CTAP-III. The Journal of Laboratory and Clinical Medicine. 1990;**116**:516-526

[47] Tai PK, Liao JF, Hossler PA, et al. Regulation of glucose transporters by connective tissue activating peptide-III isoforms. The Journal of Biological Chemistry. 1992;**267**:19579-19586

[48] Iida N, Haisa M, Igarashi A, et al. Leukocyte-derived growth factor links the PDGF and CXC chemokine families of peptides. The FASEB Journal. 1996;**10**:1336-1345

[49] Resmi KR, Krishnan LK. Proteaseaction and generation of beta-thromboglobulin-like protein followed by platelet activation. Thrombosis Research. 2002;**106**:229-236

[50] Harter L, Petersen F, Flad HD, et al. Connectivet issue-activating peptide III desensitizes chemokine receptors on neutrophils. Requirement forth proteolytic formation of the neutrophil-activating peptide 2. The Journal of Immunology. 1994;**153**:5698-5708

[51] Haskill S, Peace A, Morris J, Sporn SA, Anisowicz A, Lee SW, et al. Identification of three related human GRO gene encoding cytokine functions. USA: Proceedings of the National Academy of Sciences. Oct 1990;87(19):7732-7736. DOI: 10.1073/pnas.87.19.7732

[52] Ravindran A, Sawant KV, Sarmiento J, Navarro J, Rajarathnam K. Chemokine CXCL1 dimer is a potent agonist for the CXCR2 receptor. The Journal of Biological Chemistry. 2013;**288**(17):12244-12252

[53] Becker S, Quay J, Koren HS, Haskill JS. Constitutive and stimulated MCP-1, GRO alpha, beta, and gamma expression in human airway epithelium and bronchoalveolar macrophages. The American Journal of Physiology. 1994;**266**(3 Pt 1):L278-L286

[54] Ma K, Yang L, Shen R, Kong B, Chen W, Liang J, et al. Th17 cells regulate the production of CXCL1 in breast cancer. International Immunopharmacology. 2018;**56**:320-329

[55] Silva RL, Lopes AH, Guimarães RM, Cunha TM. CXCL1/CXCR2 signaling in pathological pain: Role in peripheral and central sensitization. Neurobiology of Disease. 2017;**105**:109-116

[56] Devalaraja RM, Nanney LB, Du J, Qian Q, Yu Y, Devalaraja MN, et al. Delayed wound healing in CXCR2 knockout mice. The Journal of Investigative Dermatology. 2000;**115**(2):234-244

[57] Reinecker HC, Loh EY, Ringler DJ, Mehta A, Rombeau JL. Mac Dermott RP. Monocyte-chemoattractant protein 1 gene expression in intestinal epithelial cells and inflammatory bowel disease mucosa. Gastroenterology 1995;**108**:40-50

[58] Chen X, Jin R, Chen R, Huang Z. Complementary action of CXCL1 and CXCL8 in the pathogenesis of gastric carcinoma. International Journal of Clinical and Experimental Pathology. 2018;**11**(2):1036-1045

[59] Hsu YL, Chen YJ, Chang WA, Jian SF, Fan HL, Wang JY, et al. Interaction between tumor-associated dendritic cells and colon cancer cells contributes to tumor progressionvia via CXCL1. International Journal of Molecular Sciences. 2018;**19**(8):2427

[60] Dhawan P, Richmond A. Role of CXCL1 in tumorigenesis of melanoma. Journal of Leukocyte Biology. 2002;**72**(1):9-18

[61] Tsai HH, Frost E, To V, Robinson S, Ffrench-Constant C, Geertman R, et al. The chemokine receptor CXCR2 controls the positioning of oligodendrocyte precursors in developing spinal cord by arresting their migration. Cell. 9 Aug 2002;**110**(3):373-383

[62] Omari KM, Lutz SE, Santambrogio L, Lira SA, Raine CS. Neuroprotection and remyelination after autoimmune demyelination in mice that inducibly overexpress CXCL1. The American Journal of Pathology. 2009;**174**(1):16476

[63] Wolpe SD, Davatelis G, Sherry B, Beutler B, Hesse DG, Nguyen HT, et al. Macrophages secrete a novel heparin-binding protein with inflammatory and neutrophil chemokinetic properties. The Journal of Experimental Medicine. 1988;**167**(2):570-581

[64] Sherry B, Tekamp-Olson P, Gallegos C, Bauer D, Davatelis G, Wolpe SD, et al. Resolution of the two components of macrophage inflammatory protein 1, and cloning and characterization of one of those components, macrophage inflammatory protein 1 beta. The Journal of Experimental Medicine. 1988;**168**(6):2251-2259

[65] Davatelis G, Wolpe SD, Sherry B, Dayer JM, Chicheportiche R, Cerami A. Macrophage inflammatory protein-1: A prostaglandin-independent endogenous pyrogen. Science. 1989;**243**(4894 Pt 1): 1066-1068

[66] Guan E, Wang J, Norcross MA. Identification of human macrophage inflammatory proteins 1alpha and 1beta as a native secreted heterodimer. The Journal of Biological Chemistry. 2001;**276**(15):12404-12409

[67] Bystry RS, Aluvihare V, Welch KA, Kallikourdis M, Betz AG. B cells and professional APC srecruitregulatory T cells via CCL4. Nature Immunology. 2001;**2**(12):1126-1132

[68] Cocchi F, DeVico AL, Garzino-Demo A, Arya SK, Gallo RC, Lusso P. Identification of RANTES, MIP-1 alpha, and MIP-1 beta as the major HIV-suppressive factor sproducedby CD8+ T cells. Science. 1995;**270**(5243):1811-1815

[69] Kamin-Lewis R, Abdelwahab SF, Trang C, Baker A, DeVico AL, Gallo RC, et al. Perforin-low memory CD8+ cells are the predominant T cells in normal humans that synthesize the beta -chemokine macrophage inflammatory protein-1 beta. Proceedings of the National Academy of Sciences of the United States of America. 2001;**98**(16):9283-9288

[70] Donlon TA, Krensky AM, Wallace MR, Collins FS, Lovett M, Clayberger C. Localization of a human T-cell-specific gene, RANTES (D17S136E), to chromosome 17q11.2-q12. Genomics. 1990;**6**(3):548-553

[71] Jumpupto Chen L, Zhang Q, Yu C, Wang F, Kong X. Functional roles of CCL5/RANTES in liver disease. Liver Research. 2020;**4**(1):28-34

[72] Appay V, Rowland-Jones SL. RANTES: A versatile and controversial chemokine. Trends in Immunology. 2001;**22**(2):83-87

[73] Maghazachi AA, Al-Aoukaty A, Schall TJ. CC chemokines induce the generation of killer cells from CD56+ cells. European Journal of Immunology. 1996;**26**(2):315-319

[74] Zeng Z, Lan T, Wei Y, Wei X. CCL5/CCR5 axis in human diseases and related treatments. Genes & Diseases. 2022;**9**(1):12-27

[75] Krensky AM, Ahn YT. Mechanisms of disease: Regulation of RANTES (CCL5) in renal disease. Nature Clinical Practice Nephrology. 2007;**3**(3):164-170

[76] Nieto M, Frade JM, Sancho D, et al. Polarization of chemokine receptors to the leading edge during lymphocyte chemotaxis. Journal of Experimental Medicine. 1997;**186**:153-158

[77] Chuluyan HE, Schall TJ, Yoshimura T, et al. IL-1 activation of endothelium supports VLA-4 (CD49d/CD29)-mediated monocyte trans

endothelial migration to C5a, MIP-1 alpha, RANTES, and PAF but inhibitsmigrationto MCP-1: A regulatory role endothelium-derived MCP-1. Journal of Leukocyte Biology. 1995;**58**:71-79

[78] Szabo MC et al. RANTES stimulation of T lymphocyte adhesion and activation: A role for LFA-1 and ICAM-3. European Journal of Immunology. 1997;**27**

[79] Bacon KB, Premack BA, Gardner P, et al. T cell signaling pathways by the chemokine RANTES. Science. 1995;**269**:1727-1730

[80] Appay V, Dunbar PR, Cerundolo V, et al. RANTES activates antigen-specific cytotoxic lymphocytes in a mitogen-like manner through cell surface aggregation. International Immunology. 2000;**12**:1173-1182

[81] Gilat D, Hershkoviz R, Mekori YA, et al. Regulation of adhesion of CD4+ T lymphocytestoin tractor heparinase-treated subendothelial extracellular matrix by diffusible or anchored RANTES and MIP-1 beta. The Journal of Immunology. 1994;**153**:4899-4906

[82] Zhang X, Chen L, Bancroft DP, et al. Crystal structure of recombinant human platelet factor 4. Biochemistry. 1994;**33**:8361-8366

[83] Baltus T, Weber KS, Johnson Z, et al. Oligomerization of RANTES is required for CCR1-mediated arrest but not CCR5-mediated transmigration of leukocytes on inflamed endothelium. Blood. 2003;**102**:1985-1988

[84] Witt DP, Lander AD. Differential binding of chemokines to glucosamine glycan subpopulations. Current Biology. 1994;**4**:394-400

[85] Hoogewerf AJ, Kuschert GS, Proudfoot AE, et al. Glycosaminoglycans mediate cell surface oligomerization of chemokines. Biochemistry. 1997;**36**:13570-13578

[86] Ali S, Palmer AC, Banerjee B, et al. Examination of the function of RANTES, MIP-1alpha, and MIP1beta following interaction with heparin-like glucosamine glycans. The Journal of Biological Chemistry. 2000;**275**:11721-11727

[87] Martin L, Blanpain C, Garnier P, et al. Structural and functional analysis of the RANTES-glycosaminoglycans interactions. Biochemistry. 2001;**40**:6303-6318

[88] Huo Y, Schober A, Forlow SB, et al. Circulating activated platelets exacerbate atherosclerosis in mice deficient in apolipoprotein E. Nature Medicine. 2003;**9**:61-67

[89] Massberg S, Brand K, Gruner S, et al. A critical role of platelet adhesion in the initiation of atherosclerotic lesion formation. Journal of Experimental Medicine. 2002;**196**:887-896

[90] Rothenbacher D, Muller-Scholze S, Herder C, et al. Differential expression of chemokines, risk of stable coronary heart disease, and correlation with established cardiovascular risk markers. Arteriosclerosis, Thrombosis, and Vascular Biology. 2006;**26**:194-199

[91] Muhlhauser I, Schernthaner G, Silberbauer K, et al. Platelet proteins (beta-TG and PF4) in atherosclerosis and related diseases. Artery. 1980;**8**:73-79

[92] Levine SP, Lindenfeld J, Ellis JB, et al. Increased plasma concentrations of platelet factor 4 in coronary artery disease: A measure of in vivo platelet activation and secretion. Circulation. 1981;**64**:626-632

[93] O'Brien JR, Etherington MD, Pashley M. Intraplatelet platelet factor 4 (IP.PF4) and the heparin-mobilisable pool of PF4 in health and atherosclerosis. Thrombosis and Haemostasis. 1984;**51**:354-357

[94] Pitsilos S, Hunt J, Mohler ER, et al. Platelet factor 4 localization in carotid atherosclerotic plaques: Correlation with clinical parameters. Thrombosis and Haemostasis. 2003;**90**:1112-1120

[95] Lambert MP, Sachais BS, Kowalska MA. Chemokines and thrombogenecity. Thrombosis and Haemostasis. 2007;**97**:722-729

[96] Rhee JS, Black M, Schubert U, et al. The functional role of blood platelet components in angiogenesis. Thrombosis and Haemostasis. 2004;**92**:394-402

[97] Brill A, Elinav H, Varon D. Differential role of platelet granular mediators in angiogenesis. Cardiovascular Research. 2004;**63**:226-235

[98] Hristov M, Zernecke A, Bidzhekov K, et al. Importance of CXC chemokine receptor 2 in the homing of human peripheral blood endothelial progenitor cells to sites of arterial injury. Circulation Research. 2009;**7**(Suppl. 1):31-34. DOI: 10.1161/01.RES.0000259043.42571.68

[99] Lasagni L, Francalanci M, Annunziato F, et al. An alternatively spliced variant of CXCR3 mediates the inhibition of endothelial cell growth induced by IP-10, Mig, and I-TAC, and acts as a functional receptor for platelet factor 4. Journal of Experimental Medicine. 2003;**197**:1537-1549

[100] Balkwill F. Cancer and the chemokine network. Nature Reviews Cancer. 2004;**4**:540-550

[101] Sallusto F, Mackay CR, Lanzavecchia A. Selective expression of the eotaxin receptor CCR3 by human T helper 2 cells. Science. 1997;**277**:2005-2007

[102] Debidda M, Williams DA, Zheng Y. Rac1 GTPase regulates cell genomic stability and senescence. The Journal of Biological Chemistry. 2006;**281**:38519-38528

[103] Acosta JC, Loghlen A, Banito A, Raguz S, Gil J. Control of senescence by CXCR2 and its ligands. Cell Cycle. 2008;**7**:2956-2919, 2959

[104] Lijnen HR. Pleiotropic functions of plasminogen activator inhibitor-1. Journal of Thrombosis and Hemostasis. 2005;**3**(1):35-45

[105] Khan SS. The central role of PAI-1 in COVID-19: Thrombosis and beyond. The American Journal of Respiratory Cell and Molecular Biology. 2021;**65**(3):238-240

Chapter 2

Alteration of Cytokines Level and Oxidative Stress Parameters in COVID-19

Marija Petrusevska, Emilija Atanasovska, Dragica Zendelovska, Aleksandar Eftimov and Katerina Spasovska

Abstract

In addition to the proinflammatory state, cytokine production, and cell death, SARS-CoV-2 infection is also associated with oxidative stress as demonstrated by increase in reactive oxygen species (ROS) levels and an alteration of antioxidant defense during the infection. Proinflammatory cytokines and chemokines play an important role in respiratory infections caused by viruses including SARS-CoV-2 by activation of the adaptive immune response. In case when the response is not controlled, it can lead to lung tissue involvement in the course of acute respiratory distress syndrome (ARDS) or can result in multiple organ failure. Oxidative stress markers show good correlation with several cytokines, which can be measured at the beginning of the disease in a primary care setting to predict the course of COVID-19.

Keywords: oxidative stress, cytokines, COVID-19, NLR, PLR

1. Introduction

Even though the pathogenesis of COVID-19 is still unclear, during the past 2 years, we can say that in general the mechanisms involved in this disease depend on several modalities with the final aim of the virus to learn how to escape the immune system of the host [1, 2]. Virus pathogenicity, comorbidities of the infected individual, and the ability of the host immune system to respond to the induced cytopathic effect have significant impact on the progress and outcome of the disease. The virus SARS-CoV-2 has so far affected more than 400 million people globally, with over 5 million deaths. The largest number of deaths, approximately 73%, is in the population over 65 years of age [3].

Cytokine storm syndrome has been widely discussed and proposed as one of the underlying etiologies of respiratory failure in patients infected with SARS-CoV-2. Cytokines present a group of polypeptides signaling molecules responsible for regulation of number of biological processes by using cell surface receptors. Proinflammatory cytokines and chemokines play an important role in respiratory infections caused by viruses including SARS-CoV-2 by activation of the adaptive immune response. When infected with influenza, an excessive amount of reactive oxygen species (ROS) is generated in several tissues including the alveolar endothelium and epithelium [4] where induced cytokines expression by activation of Toll-like receptors (TLRs) is reported to be responsible for the pathogenesis [5, 6]. Increased oxidative stress is also present during infections with human respiratory syncytial virus [7], rhinoviruses [8], and many other viruses, which has been a subject of discussion in many previously published reviews [9–16], and several other experimental studies propose that the so-called cytokine storm correlated with direct tissue injury, which afterward results in unfavorable prognosis in patients with severe COVID-19 [10]. In patients with severe COVID-19, increased levels of several cytokines, namely IL-6, IL-10, IL-2R, and TNF-α, have been reported in recently published articles on this subject [17, 18]. However, other authors suggest that more cytokines, such IL-1β, IL-1RA, IL-8, and IL-18, are also included in the pathogenesis of SARS-CoV-2 infection [10, 17, 18].

In general, patients infected with SARS-CoV-2 have either normal or reduced white blood cells (WBC) count and lymphocytopenia, and patients with severe form of the disease have additional presence of significant increase of elevated neutrophil levels, D-dimers, accompanied with continuous decrease of lymphocytes and increase of levels of certain cytokines and chemokines.

The infection with SARS-CoV-2 follows the same pathway as the innate immune response as suggested by several authors [4, 14]. That is to say, reactive oxygen species present a strong ligand and a direct mediator in the inflammasome (NLP3) trigger. Furthermore, the reactive oxygen species activate the NF-κB (nuclear factor kappa-light-chain-enhancer of activated B cells), and this further elicits the transcriptional levels of NLPR3, which are additionally enhanced by TLR and NLR (nod-like receptor) ligands. As a final outcome, this means that the ROS can increase inflammasome either directly or indirectly [12, 16]. In addition of ROS, H_2O_2 activates the NF-κB, which contributes to additional production of inflammatory cytokines [19]. Increased levels of the following cytokines: IL-6, TNF-α, IL-1β, IP-10, GCSF, MCP-1, MIP1-α/CCL3, followed by elevated blood ferritin levels, are also observed in patients infected with SARS-CoV-2 [11, 20]. Literature data has also discussed that the imbalance in the T-helper cell subsets (Th1/Th2/Th7) and regulatory T-cells also contribute to the COVID-19 pathogenesis. CD4+ T cells are divided into different subtypes based on their cytokine production, namely Th1 is producing IFN-γ, IL-2, and TNF-α, Th2: IL-4, IL-5, IL-9, IL-13, whereas the regulatory T-cells are producing TGF-β and IL-10, among others.

In this chapter, authors would like to share some experimental data, which were obtained in the last 2 years since the beginning of the COVID-19 pandemic. First, we will present results obtained by a highly standardized cytokine assay where we have measured plasma levels of IL-2, IL-4, IL-6, IL-8, IL-10, VEGF, IFN-γ, TNF-α, IL-1α, MCP-1, and EGF potentially associated as key factors with the cytokine storm syndrome in critically ill COVID-19 patients. Afterward, we have investigated which of these cytokines involved in the cytokine storm of COVID-19 show good association/correlation with the oxidative stress markers determined with fast and inexpensive photometric analytical method [20, 21]. Also we will present results in regard to the relation between the cytokines, oxidative stress markers, and the most commonly used inflammation-related biomarkers (CRP, D-dimers, PLR, NLR, and LDH) in severe form of the disease will be discussed. Additionally, the readers can obtain information for which we believe could enhance the knowledge of the altered lymphocyte subsets and their correlation with the oxidative stress markers as a tool with prognostic power to differentiate between moderate and severe COVID-19.

2. Clinical laboratory parameters abnormalities in moderate and severe COVID-19 patients

Our results obtained from 50 hospitalized COVID-19 patients who were hospitalized at the University Clinic for Infectious Diseases and Febrile Conditions, Skopje, Republic of North Macedonia, within a period of 2 months at the beginning of the pandemic were included in this study. All patients were confirmed to have SARS-CoV-2 infection by real-time reverse transcriptase–polymerase chain reaction assay from nasal and pharyngeal swab specimen. The diagnosis and classification of COVID-19 were based on the Interim Guidance for Clinical Management of COVID-19 issued by the World Health Organization. Patients with moderate form of the disease were adults from both sexes with clinical signs of pneumonia with no signs of severe pneumonia and $SpO_2 > 90\%$ on room air. In terms of the severe cases, they additionally had at least one of the following conditions: $SpO_2 < 90\%$ on room air, respiratory rate more than 30 breaths per minute, or presence of severe respiratory distress.

The work was performed by a multidisciplinary group, including clinical experts in COVID-19 management.

When admitted to hospital, patients classified with the moderate form of the disease demonstrated abnormal values for CRP, LDH, glucose, and NLR. Namely, on hospital discharge, these clinical laboratories were improved significantly, namely the mean value for C-reactive protein (CRP) was 44.1 mg/L versus 7.16 mg/L, LDH was 280.3 IU/L versus 283.4 IU/L, and neutrophil to leukocyte ratio (NLR) was 6.5 versus 5.1 at admission and discharge, respectively. Whereas patients with severe form of the disease had more prominent abnormal values of CRP, LDH, creatinine kinase (CK), ALT, AST, glucose, WBC, and NLR levels. It is worth noting that the mean values of CRP, LDH, CK, ALT, and NLR in the patients with severe COVID-19 were higher when compared with the patients with the moderate form of the disease upon admission and discharge. In general, our results are added value to data obtained by other research groups where lymphopenia can be used as reliable marker of the severity of the disease and the necessity for hospitalization. Meta-analyses by Lagunas-Rangel [22] demonstrated that patients with severe form of COVID-19 had increased NLR and decreased lymphocyte-CRP ratio when compared with moderate and mild form of the disease. Also, large number of studies, including our results, had showed that in the patients with severe form of the disease and the ones that did not survive had low platelet counts. Namely, the results from our hospitalized patients with severe COVID-19 had increased WBC, neutrophils, NLR, and platelet to lymphocyte ratio (PLR). The coagulation profile revealed elevated levels for D-dimer, prolonged PT with normal aPPT and TT. Patients with severe form of the disease that have recovered had a statistically significant difference for NLR and PLR ($p < 0.01$), and the values were decreased to 70% and 67%, respectively, when compared with the values of these parameters on admission. Additionally we have observed increase in the lymphocytes for 53%. Also, WBC decreased to 18%, neutrophils to 11%, platelets to 19%, PT to 22%, and D-dimers to 85%, but the difference was not statistically significant when compared between admission and hospital discharge ($p > 0.05$). Patients that had deterioration of their condition and died had continuous increase in WBC and neutrophils (34%), NLR (55%), decrease in lymphocytes (42%), PT (18%), and D-dimers (50%). We have only observed statistically significant difference only for the lymphocytes as parameter compared with the admission values. Reported results from meta-analysis [23] with 1779 patients demonstrated that thrombocytopenia is associated with fivefold increased risk for disease complications and death. D-dimer has also been applied as predictor for developing complications of the disease such as requiring mechanical ventilation, developing of acute respiratory distress syndrome, etc. Experiences from different clinical centers worldwide suggest that patients who had significant increase in the D-dimers should be considered for hospitalization even if there is an absence of symptoms.

All hospitalized patients whether classified with the moderate or the severe form of the disease had low partial pressure of oxygen, although it did not show statistically significant difference between the two groups. However, this observation had clinical importance indicating the need for supplemental oxygen therapy in both groups with moderate and with severe disease. The measured oxygen saturation was lower in the severe group of patients (86.26%), and this parameter had strong statistical significance, supporting the clinical indication for application of oxygen therapy. The analysis showed that in both groups, the partial pressure of carbon dioxide and pH values were within the reference values, the levels of bicarbonates were slightly increased, but without a significant difference between the patients with moderate and severe form of the disease. The base excesses were increased in the group of patients with moderate disease, and it was statistically significant, but no disturbances in acid balance ratio occurred within any of the group. Similar results were reported by Mumoli et al. [24] at hospital admission of 88 COVID-19 patients, as well as by Doaei et al. [25], who measured similar values of the blood gas parameters in critically ill patients with COVID-19 to those reported in our study. On the other hand, Deniz et al. [26] detected a mild increase of pH and bicarbonate and relatively low pCO_2 in COVID-19 patients compared with non-COVID-19 individuals.

3. Cytokine alteration and oxidative stress in patients with severe COVID-19

Cytokines as well as chemokines and growth factors together with the lipid metabolites present one of the key players in the immune cell function and their differentiation, which on the other hand, means that following a dysregulation in the process, various diseases can arise [11–13, 16]. In this chapter, we present some of our recently obtained results as an add-on to the clinical and scientific evidences in terms of oxidative stress, which is increased in patients with severe COVID-19. We have also obtained results that the measured oxidative stress parameters in these patients show a good correlation with the level of cytokines and with some of the commonly used laboratory biomarkers. This was a case–control pilot study focused primarily on the possibility to apply the oxidative stress parameters (d-ROM, PAT, and OS index) measured on a spectrophotometric system as a fast and low-cost prognostic tool for disease progression and potentially predict the outcome of COVID-19 in patients. Abnormal levels of several cytokines involved in the adaptive immunity (IL-2, IL-4) or proinflammatory cytokines and interleukins (IFNs, IL-1, IL-6, IL-10 IL-17, and TNF-α) were reported [11, 17, 27, 28] mainly as results from retrospective studies and reviews.

In our case–control pilot study, we have measured the levels of 11 cytokines (IL-2, IL-4, IL-6, IL-8, IL-10, VEGF, IFN-γ, TNF-α, IL-1α, MCP-1, and EGF) in 14 critically ill patients. Afterward, these values were compared with the levels of the same cytokines in individuals that were not infected with the SARS-CoV-2 virus. By using t-test, we have obtained statistically significant increase ($p < 0.05$), which was observed in regard to the levels of IL-6, IL-8, IL-10, VEGF, MCP-1, and EGF in the SARS-CoV-2 patients. Additionally, the levels of IL-2, IFN-γ, TNF-α, and IL-1α were also increased, but, however, this difference was not considered to be significant when compared with the noninfected individuals ($p > 0.05$) [20, 21]. Important finding of our case–control pilot study was that the oxidative stress parameters, d-ROM (448.8 U.Carr), OS index (107.7), and PAT (3048 U.Corr), were significantly higher ($p < 0.05$) in the infected patients when compared with those not infected. Additionally, we have also explored the correlation among the abovementioned cytokines (IL-2, IL-4, IL-6, IL-8, IL-10, VEGF, IFN-γ, TNF-α, IL-1α, MCP-1, and EGF), the oxidative stress parameters (d-ROM, PAT, OSI), and some of the commonly used laboratory parameters (CRP, LDH, PLR, D-dimer, and NLR) by using the Spearman r calculation. These calculations were generated as heat map, and we have obtained a significant positive correlation between all investigated cytokines and the parameters of the oxidative stress (d-ROM, PAT, and OSI), except between IL-10 and the total antioxidant capacity, PAT, for which a negative a correlation was obtained. According to the performed analysis, we have obtained a nonsignificant correlation between OS index and the IL-8 ($p = 0.8552$) and also between d-ROM and VEGF ($p = 0.999$). The strongest correlation was demonstrated in the case of IL-6, which was estimated as significant with all of the markers of the oxidative stress, d-ROM ($r = 0.9725$, $p < 0.01$), PAT ($r = 0.5000$, $p < 0.01$), and the oxidative stress index ($r = 0.9593$, $p < 0.05$). Also, our results present evidence for similar behavior between IFN-γ and d-ROM ($r = 0.4006$, $p < 0.01$), PAT ($r = 0.6030$, $p < 0.01$) and the oxidative stress index ($r = 0.4298$, $p < 0.05$). Additionally, statistically we have investigated the correlation between the commonly used inflammation biomarker, CRP, and the levels of the investigated cytokines. In this case we have observed a strongest correlation with several of the investigated cytokines, namely with IL-6, IL-8, MCP-1, and IFN-γ. Furthermore, in terms of correlation, a strong correlation was obtained between the investigated inflammatory cytokines IL-2, IL-4, IL-6, IL-8, IL-10, VEGF, IFN-γ, TNF-α, IL-1α, and MCP-1, except between IL-6 and EGF where negative correlation was obtained. It has been demonstrated that the high expression of IL-6 in COVID-19 patients can hasten the inflammatory process, which can on the other hand contribute to the cytokine storm and can contribute to disease worsening. The importance of the IL-6 has been potentiated by using tociluzimab, a monoclonal antibody, which can block the receptor of IL-6 for which it has been reported that the cytokine hyperproduction has been reversed [10].

The research performed by our study group has contributed to the evidences that more than a few cytokines and other clinically relevant biomarkers were significantly increased in patients with severe form COVID-19 in comparison to the individuals that were not infected by the virus, which was followed with coagulopathy as determined by worsening of the platelet-related parameters (PLR, D-dimers, and IL-6) and the increased levels of MCP-1 as thrombosis-related indicator. MCP-1 levels were estimated to be much higher in critical hospitalized patients in the ICU accompanied by decrease of platelet count in patients that do not survive, which was reported also by Huang et al. [27]. In our case–control pilot study, unfortunately all patients had severe form of COVID-19 and all of them had died during hospitalization. Besides, in our study we have witnessed that the levels of several of the investigated cytokines had been increased more than 10 times above the levels of the noninfected, which we have considered as a baseline. It is worth noticing that the significant increase of the vascular endothelial growth factor (VEGF) levels more than 10 times can be related to the essential VEGF role in the endothelial cell activation by binding to cell surface

VEGF receptors. This increase was considered to be statistically significant ($p < 0.01$). The upregulation of VEGF was also observed in several other viral infections, and also it has been investigated as a potential targeted therapy in viral diseases [28]. Higher levels of VEGF in hospitalized COVID-19 patients were also reported in a published study by Huang et al. [27]. The cytokine profile in COVID-19 pregnant women was investigated in a published study by Tanacan et al. [29], in which this research group has reported significantly higher values for IFN-γ and IL-6 with lower values of IL-2, IL-10, and IL-17. This situation was especially pronounced in those patients who had complications such as miscarriage and preterm delivery [29].

From the above presented results and the available literature data, we can conclude that the strong correlation between the investigated cytokines including chemokines and growth factors (IL-2, IL-4, IL-6, IL-8, IL-10, VEGF, IFN-γ, TNF-α, IL-1α, MCP-1, and EGF), the oxidative stress parameters (d-ROM, OSI, and PAT), and some of the commonly used clinical markers (CRP, D-dimers, NLR, PLR) is in agreement with the proposed "cytokine storm" as potential mechanism of the SARS-CoV-2 infection. In general, the so-called "cytokine storm" occurs when large numbers of white blood cells (leukocytes) are activated followed by release of a high concentration of proinflammatory cytokines, primarily IL-6, IL-10, IFN, MPC-1, IL-1, IL-2, and IL-8. Mostly, SARS-CoV-2 infection is related with increased oxidative stress, the proinflammatory state, cytokine production, and cell death confirmed by increase in the levels of the ROS and an alteration of antioxidant defense during the infection [15, 21, 30]. This interplay is depicted in **Figure 1**.

As proposed previously [11, 27, 30], RNA viruses trigger the oxidative stress by disturbing the pro-antioxidant–antioxidant balance for which we believe that the same mechanism exists also during the infection with SARS-CoV-2. During the last 2 years, we have performed several investigations with COVID-19 patients and we have managed to demonstrate increased oxidative stress and increased cytokines levels in these patients. Actually, we have observed significantly higher level of the markers of the oxidative stress (d-ROM and OS index values) and decreased total antioxidant capacity (PAT) in the COVID-19 patients when compared with disease-free individuals, supporting the theory that viral infection will increase the oxidative stress in the organism and further complicate the course of the disease (**Figure 1**).

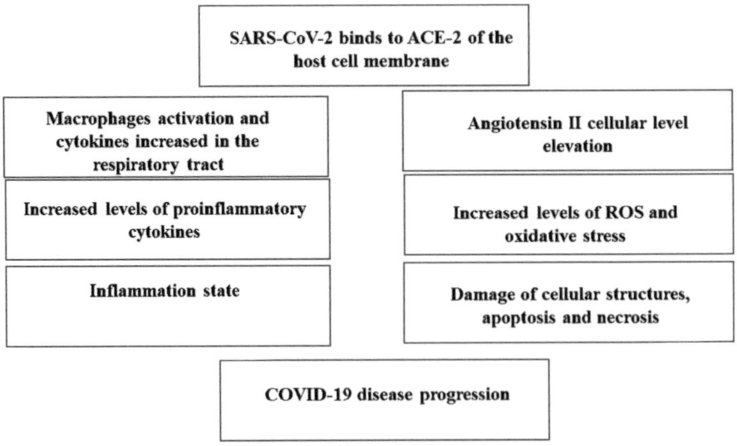

Figure 1.
Interplay of factors in COVID-19.

While we consider that the oxidative stress index value presents a significant parameter, which can be used in clinical practice for patient classification, we can also try in future activities, to implement a certain principle of supplementation with antioxidants especially when there is relevant knowledge for more than a few vitamins (vitamin C, vitamin D, selenium, quercetin, and other polyphenols) with proven antioxidant, anti-inflammatory, and even antiviral capacity [31]. Also, information on several clinical trials for potential therapy in COVID-19, such as tocilizumab, a monoclonal antibody IL-6 receptor antagonist; sarilumab – IL-6 antagonist; and anakinra, a recombinant IL-1 receptor antagonist against selected cytokines is in progress, some of which have already been published [31–38]. Even though, data on their efficacy are evolving and additional studies are needed for those treatments to be administered routinely in COVID-19.

3.1 Alteration of oxidative stress markers, inflammatory cytokines, and lymphocyte subsets in moderate and severe COVID-19

In regard to the pathophysiology of the virus, SARS-CoV-2 enters into the cell through the angiotensin-converting enzyme-2 (ACE-2), mainly through the Toll-like receptor-7 (TLR-7) that is present in the endosomes (**Figure 1**). The activation of this receptor requires production of IL-6, IL-12, and TNF-a enabling cytotoxic CD8+ T cells generation. This process results in further formation of antigen-specific B- cells and antibodies production through CD4+ helper T cells [39]. The suppression of the immune response can occur as negative reaction from immune activation and that is one of the reasons we have additionally performed analysis on several lymphocyte subsets in one of our studies. Th1 cells, natural killer cells (NK cells), and CD8+ T cells are main sources of INF-γ, whose production has been increased in COVID-19 patients. The increase of this cytokine (INF-γ) triggers Th1 response to eradicate the viral infect as one of the human immune strategies. Results reported since the beginning of the COVID-19 pandemic stated that more than a half of the patients had low lymphocytes number, and moreover pathological findings suggested that the patients who died of COVID-19 had significant increase of the concentration of proinflammatory CCR6+ Th17 within the CD4+ cells [2]. This point toward that overactivation of T-cells in combination with high cytotoxicity of CD8 T cells is partially responsible for severe immune injuries in COVID-19 patients. Several reported clinical studies indicate that low CD8+ T cells counts and high neutrophil-to-lymphocyte ratios (NLR) are associated with increased risk for disease severity and mortality in critically ill COVID-19 patients [40–43]. Namely, the high ratio between the neutrophils and leukocytes in COVID-19 patients leads to redox imbalance as a result of increased reactive oxygen species (ROS) production. In addition, the activation of macrophages and the polymorphonuclear cells also has effect on the oxidative damage to the tissues, and they can lead to organ failure. Patients with moderate form of COVID-19 had lower values of the measured concentration of free radicals (d-ROMs) and hence lower oxidative stress index (OSI) when compared with the patients classified with the severe form of the disease ($p < 0.01$). Additionally, the moderate group had increased total antioxidant capacity (PAT) in comparison with the severe group of patients, but however, this difference did not reach a statistical significance ($p > 0.05$). In our previous study, d-ROM (concentration of free radicals) and OSI demonstrated a good correlation with IL-6 and VEGF out of the 11 screened cytokines as predictors of disease worsening in severe COVID-19 patients [21]. In this study we have used them as parameters to potentially predict the so-called "cytokine storm," and herein we report a statistically significant difference of IL-6 and VEGF levels between the two groups of patients (moderate vs. severe) ($p < 0.01$).

In terms of the alteration of the lymphocyte subsets, we have observed decreased levels of leukocytes and its subsets, namely CD4+, CD8+, CD3+, NK cells, and the absolute number of CD8 ($p < 0.05$). Also, CD19+ and CD45+ were decreased in the severe group in comparison to the moderate group of patients, but the difference did not reach a statistical significance ($p > 0.05$), presumably due to small sample size. These results are shown in **Table 1**.

Moreover, we have also investigated the correlation by calculation of the Spearman r among all investigated parameters in both groups separately. We have demonstrated that in the moderate group, a good correlation between the levels of IL-6 and VEGF and NK cells was obtained (for IL-6, $r = 0.6973$, $p < 0.05$; and for VEGF, $r = 0.6498$, $p < 0.05$), whereas in the severe group only these cytokines correlated with CD45+ (for IL-6, $r = 0.5610$, $p < 0.05$; and for VEGF, $r = 0.5462$, $p < 0.05$). The results from our investigation had shown that both parameters CD45+ and the oxidative stress index can be considered as an applicable diagnostic standard in distinguishing severe form of the disease and disease complications.

The oxidative stress index can be used as potential marker with a diagnostic value since we have obtained very high values in patients with severe COVID-19 as well as OSI demonstrated a good correlation with IL-6, CD45+, CD4+, and absolute number of CD8 cells. In the severe group, we have observed a high level of the proinflammatory cytokine IL-6, which probably contributes to the T lymphocyte deficiency (CD4+, CD8+, CD3+) [17, 28]. Several authors have reported that during viral infections with the virus, CD4+ T lymphocytes are activated into T helper cells, and then they secret proinflammatory cytokines as IL-6. The activated immune cells enter into pulmonary circulation and can lead to serious lung injury [17, 21, 28]. In a study of Mudd et al. [44], gene expressions differences between influenza and COVID-19 patients were investigated. Namely, in comparison to the influenza state, INF-γ and IFN-α response pathways were downregulated within the patients with COVID-19 in terms of B-cells, CD8+ T cells, regulatory T cells, plasma blasts, and monocyte subsets. Hence, because of this interplay between the defect in the host immunity and the increased cytokines level, the evaluation of the adaptive and the innate immunity of the COVID-19 patient might be useful in terms of immunomodulatory therapies.

Subset	Severe COVID-19 patients mean ± SEM (n = 16)	Moderate COVID-19 patients mean ± SEM (n = 19)	p (t-test)
Leukocytes	12.28 ± 1.055	7.637 ± 0.8581	**0.0016**
CD4+	0.1711 ± 0.0184	0.6765 ± 0.0653	**0.0001**
CD8+	0.1034 ± 0.0191	0.3157 ± 0.0332	**0.0001**
NK	0.06225 ± 0.01656	0.1498 ± 0.02519	**0.0087**
Absolute CD8	99.63 ± 10.78	317.4 ± 32.42	**0.0001**
CD45+	0.7382 ± 0.1188	1.266 ± 0.1428	0.7594
CD19+	0.1852 ± 0.0495	0.2686 ± 0.0861	0.4296
CD3+	0.4038 ± 0.0629	0.9057 ± 0.0114	**0.0009**

Table 1.
Alteration in lymphocytes subsets in severe and moderate COVID-19 patients on hospital admission. Results are expressed as mean ± SEM.

4. Conclusion

Upon available data and the results from our clinical experience in the last 2 years of the pandemic, we can conclude that most probably the interplay between the defect in the host immunity and the cytokines hyperproduction is an important factor in the immunopathology of the SARS-CoV-2 infection. Our experience demonstrated that the levels of certain cytokines, namely IL-6, IL-8, IL-10, VEGF, MCP-1, and EGF, were significantly increased in the critically ill COVID-19 patients. Moreover, we have proved a good correlation of the increased level of IL-6 with the oxidative stress index, which can be considered as evidence that the cytokine storm syndrome lies as an immune-pathogenesis during SARS-CoV-2 infection.

Acknowledgements

Authors would like to thank Nikola Labachevski for critical reading of the manuscript and to the technical staff at the Clinic for Infectious Diseases and Febrile Conditions.

Conflict of interest

The authors declare no conflict of interest.

Author details

Marija Petrusevska[1*], Emilija Atanasovska[1], Dragica Zendelovska[1], Aleksandar Eftimov[2] and Katerina Spasovska[3]

1 Faculty of Medicine, Institute of Preclinical and Clinical Pharmacology and Toxicology, University of Ss Cyril and Methodius, Skopje, Republic of North Macedonia

2 Faculty of Medicine, Institute of Pathology, University of Ss Cyril and Methodius, Skopje, Republic of North Macedonia

3 University Clinic for Infectious Diseases and Febrile Conditions, Skopje, Republic of North Macedonia

*Address all correspondence to: marija.petrusevska@medf.ukim.edu.mk

References

[1] World Health Organization. Naming the Coronavirus Disease (COVID-19) and the Virus that Causes it. Geneva, Switzerland: WHO; 2020. Available from: https://www.who.int/emergencies/diseases/novel-coronavirus-2019/technical-guidance/naming-the-coronavirus-disease-(covid-2019)-and-the-virus-that-causes-it

[2] Xu Z, Shi L, Wang Y, Zhang J, Huang L, Zhang C, et al. Pathological findings of COVID-19 associated with acute respiratory distress syndrome. The Lancet Respiratory Medicine. 2020;**8**(4):420-422

[3] Wang D, Hu B, Hu C, Zhu F, Liu X, Zhang J, et al. Clinical characteristics of 138 hospitalized patients with 2019 novel coronavirus-infected pneumonia in Wuhan, China. Journal of the American Medical Association. 2020;**323**(11):1061-1069

[4] Chernyak BV, Popova EN, Prikhodko AS, Grebenchikov OA, Zinovkina LA, Zinovkin RA. COVID-19 and oxidative stress. Biochemistry (Mosc). 2020;**85**(12):1543-1553

[5] Iwasaki A, Pillai PS. Innate immunity to influenza virus infection. Nature Reviews. Immunology. 2014;**14**:315-328

[6] Hui DSC, Zumla A. Severe acute respiratory syndrome: Historical, epidemiologic, and clinical features. Infectious Disease Clinics of North America. 2019;**33**(4):869-889

[7] Menzel M, Ramu S, Calvén J, Olejnicka B, Sverrild A, Porsbjerg C, et al. Oxidative stress attenuates TLR3 responsiveness and impairs anti-viral mechanisms in bronchial epithelial cells from COPD and asthma patients. Frontiers in Immunology. 2019;**10**:2765

[8] Kaul P, Biagioli MC, Singh I, Turner RB. Rhinovirus-induced oxidative stress and interleukin-8 elaboration involves p47-phox but is independent of attachment to intercellular adhesion molecule-1 and viral replication. The Journal of Infectious Diseases. 2000;**181**(6):1885-1890

[9] Khomich OA, Kochetkov SN, Bartosch B, Ivanov AV. Redox biology of respiratory viral infections. Viruses. 2018;**10**(8):392

[10] Costela-Ruiz VJ, Illescas-Montes R, Puerta-Puerta JM, Ruiz C, Melguizo-Rodríguez L. SARS-CoV-2 infection: The role of cytokines in COVID-19 disease. Cytokine & Growth Factor Reviews. 2020;**54**:62-75

[11] Cecchini R, Cecchini AL. SARS-CoV-2 infection pathogenesis is related to oxidative stress as a response to aggression. Medical Hypotheses. 2020;**143**:110102

[12] Chen Y, Zhou Z, Min W. Mitochondria, oxidative stress and innate immunity. Frontiers in Physiology. 2018;**18**(9):1487

[13] Li H, Liu L, Zhang D, Xu J, Dai H, Tang N, et al. SARS-CoV-2 and viral sepsis: Observations and hypotheses. Lancet. 2020;**395**(10235):1517-1520

[14] van den Brand JM, Haagmans BL, van Riel D, Osterhaus AD, Kuiken T. The pathology and pathogenesis of experimental severe acute respiratory syndrome and influenza in animal models. Journal of Comparative Pathology. 2014;**151**:83-112

[15] Chen K, Bao Z, Tang P, Gong W, Yoshimura T, Wang JM. Chemokines in homeostasis and diseases. Cellular & Molecular Immunology. 2018;**15**(4):324-334

[16] Chen Y, Wang J, Liu C, Su L, Zhang D, Fan J, et al. IP-10 and MCP-1 as biomarkers associated with disease severity of COVID-19. Molecular Medicine. 2020;**26**:97

[17] Han H, Ma Q, Li C, Liu R, Zhao L, Wang W, et al. Profiling serum cytokines in COVID-19 patients reveals IL-6 and IL-10 are disease severity predictors. Emerging Microbes & Infections. 2020;**9**(1):1123-1130

[18] Takada Y, Mukhopadhyay A, Kundu GC, et al. Hydrogen peroxide activates NF-KB through tyrosine phosphorylation of IκBα and serine phosphorylation of p65. The Journal of Biological Chemistry. 2003;**278**(26):24233-24241

[19] Zafer MM, El-Mahallawy HA, Ashour HM. Severe COVID-19 and Sepsis: Immune pathogenesis and laboratory markers. Microorganisms. 2021;**9**(1):159

[20] Zendelovska D, Atanasovska E, Petrushevska M, Spasovska K, Stevanovikj M, Demiri I, et al. Evaluation of oxidative stress markers in hospitalized patients with moderate and severe COVID-19. Romanian Journal of Internal Medicine. 2021;**59**(4):375-383. DOI: 10.2478/rjim-2021-0014

[21] Petrushevska M, Zendelovska D, Atanasovska E, Eftimov A, Spasovska K. Presentation of cytokine profile in relation to oxidative stress parameters in patients with severe COVID-19: A case-control pilot study. F1000Res. 2021;**10**:719. DOI: 10.12688/f1000research.55166.2

[22] Lagunas-Rangel FA. Neutrophil-to-lymphocyte ratio and lymphocyte-to-C-reactive protein ratio in patients with severe coronavirus disease 2019 (COVID-19): A meta-analysis. Journal of Medical Virology. 2020;**92**(10):1733-1734. DOI: 10.1002/jmv.25819

[23] Lippi G, Plebani M, Henry BM. Thrombocytopenia is associated with severe coronavirus disease 2019 (COVID-19) infections: A meta-analysis. Clinica Chimica Acta. 2020;**506**:145-148. DOI: 10.1016/j.cca.2020.03.022

[24] Mumoli N, Bonaventura A, Colombo A, Vecchié A, Cei M, Vitale J, et al. Lung function and symptoms in post–COVID-19 patients: A single-Center experience. Mayo Clinic Proceedings: Innovations, Quality & Outcomes. 2021;**5**(5):907-915

[25] Doaei S, Gholami S, Rastgoo S, Gholamalizadeh M, Bourbour F, Bagheri SE, et al. The effect of omega-3 fatty acid supplementation on clinical and biochemical parameters of critically ill patients with COVID-19: A randomized clinical trial. Journal of Translational Medicine. 2021;**19**:128

[26] Deniz S, Uysal TK, Capasso C, Supuran CT, Guler OO. Is carbonic anhydrase inhibition useful as a complementary therapy of Covid-19 infection? Journal of Enzyme Inhibition and Medicinal Chemistry. 2021;**36**(1):1230-1235

[27] Huang C, Wang Y, Li X, Ren L, Zhao J, Hu Y, et al. Clinical features of patients infected with 2019 novel coronavirus in Wuhan, China. Lancet. 2020;**395**:497-506

[28] Alkharsah KR. VEGF upregulation in viral infections and its possible therapeutic implications. International Journal of Molecular Sciences. 2018;**19**(6):1642

[29] Tanacan A, Yazihan N, Ahmet Erol A, Taner Anuk A, Yucel Yetiskin FD, Biriken D, et al. The impact of COVID-19 infection on the cytokine profile of pregnant women: A prospective case-control study. Cytokine. 2021;**140**:155431

[30] Liu J, Li S, Liu J, Liang B, Wang X, Wang H, et al. Longitudinal characteristics of lymphocyte responses and cytokine profiles in the peripheral blood of SARS-CoV-2 infected patients. eBioMedicine. 2020;**55**:102763

[31] de Las HN, Martín Giménez VM, Ferder L, Manucha W, Lahera V. Implications of oxidative stress and potential role of mitochondrial dysfunction in COVID-19: Therapeutic effects of vitamin D. Antioxidants. 2020;**9**(9):897

[32] Polonikov A. Endogenous deficiency of glutathione as the Most likely cause of serious manifestations and death in COVID-19 patients. ACS Infectious Diseases. 2020;**6**(7):1558-1562

[33] Cekerevac I, Nikolic Turnic T, Draginic N, Andjic M, Zivkovic V, Simovic S, et al. Predicting severity and Intrahospital mortality in COVID-19: The place and role of oxidative stress. Oxidative Medicine and Cellular Longevity. 2021;**2021**:6615787

[34] Lammi C, Arnolodi A. Food derived antioxidants and COVID-19. Journal of Food Biochemistry. 2021;**45**(1):e13557

[35] Ntyonga-Pono M. COVID-19 infection and oxidative stress: An under-explored approach for prevention and treatment? The Pan African Medical Journal. 2020;**35**(Suppl 2):12

[36] Bogdanic N, Lj L, Begovac J. Coronavirus disease 19 (COVID-19) pharmacologic treatment: Where are we now? Infektološki glasnik. 2020;**40**(4):134-147

[37] Malgie J, Schoones JW, Pijls BG. Decreased mortality in coronavirus disease 2019 patients treated with tocilizumab: A rapid systematic review and meta-analysis of observational studies. Clinical Infectious Diseases. 2021;**72**(11):e742-e749

[38] Huet T, Beaussier H, Voisin O, et al. Anakinra for severe forms of COVID-19: A cohort study. The Lancet Rheumatology. 2020;**2**(7):e393-e400

[39] Ahmadpoor P, Rostaing L. Why the immune system fails to mount an adaptive immune response to a COVID-19 infection. Transplant International. 2020;**33**(7):824-825. DOI: 10.1111/tri.13611

[40] Ramljak D, Vukoja M, Curlin M, et al. Early response of CD8+ T cells in COVID-19 patients. Journal of Personalized Medicine. 2021;**11**(12): 1291

[41] Zuani D, Laznickova P, Tomaskova V, et al. High CD4-to-CD8 ratio identifies an at-risk population susceptible to lethal COVID-19. Scandinavian Journal of Immunology. 2022;**95**(3):e1312

[42] Pallotto C, Suardi LR, Esperti S, et al. Increased CD4/CD8 ratio as a risk factor for critical illness in coronavirus disease 2019 (COVID-19): A retrospective multicentre study. Infectious Diseases. 2020;**52**(9):675-677

[43] Zhang H, Wu T. CD4+T, CD8+T counts and severe COVID-19:

A meta-analysis. The Journal of Infection. 2020;**81**(3):e82-e84

[44] Mudd PA, Crawford JC, Turner JS, Souquette A, Reynolds D, et al. Distinct inflammatory profiles distinguish COVID-19 from influenza with limited contributions from cytokine storm. Science Advances. 2020;**6**(50):eabe3024. DOI: 10.1126/sciadv.abe3024

Chapter 3

Chemokines at the Crossroad of Diabetes-Tuberculosis Synergy

Vivekanandhan Aravindhan and Srinivasan Yuvaraj

Abstract

The epidemic increase in diabetes mellitus (DM) is taking place in the world were one third of the population is latently infected with tuberculosis (TB). DM, as a chronic metabolic disease, weakens the immune system and increases the risk of *Mycobacterium tuberculosis* (*M.tb*) infection. In those who are already latently infected, it increases the risk of reactivation. This is called DM-TB synergy. While the role of immune cells and cytokines has been well studied in DM-TB synergy, the role played by chemokines is largely unrecognized. Chemokines are low molecular weight proteins that are rapidly secreted by both immune and non-immune cells and guide the directorial migration of these cells. Impairment in chemokine secretion or signaling can lead to delayed immune response and can mediate DM-TB synergy. This chapter describes the role played by various chemokines and their receptors in DM-TB synergy.

Keywords: diabetes, latent tuberculosis, chemokines, chemokine receptors, insulin resistance, inflammation, immunity

1. Introduction

1.1 Diabetes-tuberculosis synergy

TB, caused by *Mycobacterium tuberculosis* (*M.tb*) infection, is estimated to affect one third of the world's population. The majority of infected individuals develop asymptomatic latent TB (LTB), while ~5–10% of these individuals will progress to active pulmonary TB (ATB) [1]. The long treatment regimen, the relative inefficacy of the BCG vaccine, in addition to increased drug resistance, leads to a rapid surge in TB cases and made WHO declare TB as a global emergency in 1993 [2]. Despite the decline in the mortality rate of ATB since 2000, TB is ranked as one of the leading causes of death [1]. In 2015, there were an estimated 10.4 million incident TB cases across the world [1]. The "End TB Strategy" commenced by the World Health Organization (WHO) in 2016, aims to terminate the global TB epidemic by 2035 [1]. Targets set in this strategy include 90% reduction in TB deaths and an 80% reduction in TB incidence by 2030 [1]. The growing epidemic of DM is predicted to become one of the major global health challenges. The number of DM subjects is projected to rise from 415 million in 2015, to 642 million by 2040 [3]. It is well established by global epidemiological studies that DM patients are highly susceptible to TB due to impaired immunity. Recent studies indicate that TB patients with DM have higher bacillary load in the sputum, delayed sputum conversion and higher rates of multidrug-resistant infection [4]. Recent epidemiological surveys have also clearly shown the chance of DM-TB nexus in near future [4]. However, the exact mechanism of DM-TB synergy is yet to be fully deciphered. Inflammation has long been identified as a common denominator of both DM and TB, which however differs in both the disease conditions [5, 6]. The systemic low-grade inflammation manifested in DM is non-protective in nature and impairs anti-TB immunity [6]. While both cytokines and chemokines play an important role in anti-TB immunity, compared to cytokines, chemokines are poorly studied in DM-TB nexus [5].

1.2 Chemokines

For sustenance of life, the right cell has to be there, at the right time, at the right place, which is called as spatio-temporal regulation. This fundamental life process is coordinated by a family of highly conserved, small proteins (8–12 kDa) called chemokines. They are best known for their ability to stimulate the directorial migration of cells, most notably immune cells. However, recently it has been demonstrated that these chemokines are also involved in the migration and organization of all body cells at some point of time. Consequently, chemokines play a central role in the overall development and homeostasis, and specifically in immune responses and inflammation. The first biologically active chemokines were discovered in the late 1980s and early 1990s, based on the purification and characterization of leukocyte chemoattractant activity, present in the culture supernatants of human leukocytes, stimulated with bacterial endotoxins [7]. First among chemokines which was isolated and characterized was monocyte chemotactic protein 1 (MCP-1), which was later termed as CCL2.

1.2.1 Nomenclature

Chemokines are classified, based on their amino acid composition, specifically on the presence of a conserved tetra-cysteine motif. Variation in the precise configuration of the two cysteines near the N terminus allows chemokines to be classified into four subfamilies: CC, CXC, CX3C, and XC. In CC chemokines, these cysteines are juxtaposed directly, while CXC chemokines hold a single variable amino acid between them. The sole CX3C chemokine has three amino acids between these two cysteines, while XC chemokines, of which there are two, lack the first and the third cysteines of the motif. Although chemokines were originally named according to specific functions, a systematic nomenclature was introduced in 2000 that includes a subfamily designation (i.e., CC, CXC, CX3C, or XC), followed by the letter L (denoting 'ligand'), and then a number according to the chronology of discovery [8, 9]. To date, the official nomenclature accounts for more than 48 chemokines in humans and includes 28 CCL, 17 CXCL, 1 CX3CL and 2 XCL members. Chemokines can be induced by diverse stimuli. Apart from chemotaxis and cell adhesion they also play an important role in cellular activation, proliferation, maturation, differentiation, apoptosis, malignant transformation, and dissemination, depending upon the cell type.

1.2.2 Chemokine receptors

Chemokines signal through cell surface receptors which are seven transmembrane G protein coupled serpentine receptors (7TMGPCR) present on several cell types. The human chemokine receptor system at present consists of 20 members. The GPCR family ranks the most diverse class of cell-surface receptors. These receptors are broadly classified into conventional chemokine receptors (cCKRs) and atypical chemokine receptors (aCKRs). cCKRs are generally specific for a single chemokine family. There are 10 CC chemokine receptors, 6 CXC chemokine receptors, and 1 receptor for C and CX3C chemokines (Totally 18). The chemokine- receptor family members show bidirectional promiscuity, meaning some chemokines can bind to more than one receptor and some receptors can bind to more than one chemokine. But conventionally, promiscuity is class restricted. The immediate consequence of receptor binding is a change in the cellular cytoskeleton which results in polarization and directional migration of the cells, up the chemokine gradient.

2. Chemokines in diabetes

Molecular links between obesity, insulin resistance (IR) and DM remain incompletely understood but may include chronic inflammation, particularly in adipose tissue (AT). IR causes rapid exhaustion of pancreatic β cells due to increased insulin demand, leading to insulin deficiency. Thus, chronic DM is depicted by combined insulin deficiency (ID) and IR progressing to hyperglycemia, eventually leading to endothelial dysfunction. Endothelial dysfunction is the root cause of microvascular and macrovascular complications, accounting for significant morbidity and mortality among DM patients. Chemokines are shown to play an important role in both IR and pancreatic β cell loss.

2.1 Chemokines in adipose inflammation and insulin resistance

The role of chemokines in adipose tissue inflammation and insulin resistance is briefly illustrated in **Figure 1**. IR is a metabolic complication in which the three major insulin-sensitive organs namely adipose tissue, skeletal muscle and liver become less responsive to insulin action. It is now well established that IR predominantly starts in adipose tissue and then spills over to other insulin-sensitive organs. The primary function of insulin in adipose tissue is to increase glucose uptake, glycolysis and increased production of acetyl-CoA, which is then converted to lipids and stored as lipid droplets. Under conditions of IR, there is a decreased glucose uptake, reduced glycolysis and acetyl-CoA production, decreased lipid synthesis and storage and increased release of free fatty acids (FFAs) [8]. FFAs undergo oxidation and give rise to fatty acid peroxides (oxFFAs), which then bind to cell surface inflammatory receptors like TLRs and trigger inflammation. Animal and human studies have identified white adipose tissue (WAT) as the primary site where obesity-related chronic inflammation is initiated and exacerbated. oxFFAs bind to TLR in presence of fetuin A and brings about the activation of adipocytes which results in the secretion of pro-inflammatory cytokines like TNF-α, IL-6 and IL-1β and chemokines like MCP-1. MCP-1 promotes macrophage infiltration, while TNF-α, IL-6 and IL-1β bring about polarization of those macrophages into M1 phenotype. Infiltrating macrophages in turn secrete several pro-inflammatory cytokines and chemokines, which act on adipocytes impairing insulin action. Macrophages are the most important cell type in mediating AT inflammation. The pro-inflammatory mediators secreted by the macrophages have local effects on adipocytes and resident macrophages and also enter circulation, where they affect the skeletal muscle and liver.

The role for chemokines in the regulation of adipose tissue metabolism was suggested during early 2000. MCP-1, IL-8 and MIP-1α were the earliest known

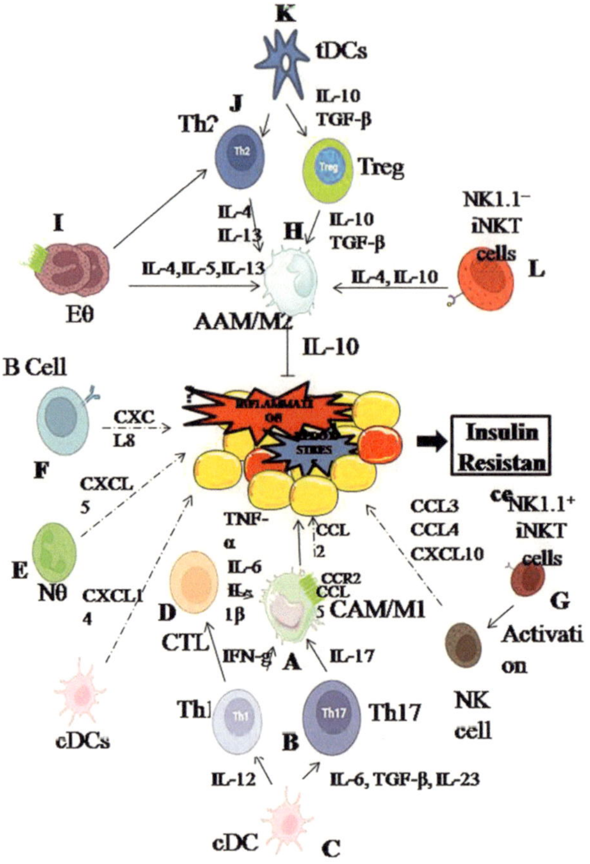

Figure 1.
Chemokines in adipose inflammation and insulin resistance. Two immune circuits plays a key role in adipose tissue inflammation, one promotes inflammation while other dampens the inflammation. A. CCR2+ Classically activated macrophages (CAM)/M1 macrophages are recruited into adipose tissue under the influence of MCP-1 secreted by the adipose and induce inflammation which is the earliest event in AT inflammation. B. Th1 & Th17 cells release IFN-gamma and IL-17 respectively which enhances M1 polarization. C. Conventional dendritic cells (cDCs) secrete IL-12 (which induces Th1 polarization) and IL-6, TGF-Beta and IL-23 (which induces Th17 polarization). cDCs also migrates to the site of inflammation through CXCL14. D. Cytotoxic T lymphocytes (CTLs) act synergistically with Th1 and Th17 cells in bringing about M1 polarization E. Neutrophils migrate to site of AT through the release of CXCL5. F. B-cells enter AT through the release of CXCL8. G. NK1.1^{+} cells upon interaction with adipocyte activate the NK cells, which then migrate to the inflammation site through the action of CCL3, CCL4 and CXCL10. H. Alternatively activated macrophages (AAM/M2) antagonize M1 macrophages through the release of IL-10. M2 macrophages are activated by I. Eosinophils by releasing IL-4, IL-5, IL13. J. Tolerogenic DCs (tDCs) polarize Tregs through the release of IL-10 and TGF-beta. L. iNKT cells release IL-4 and IL-10 which enhances the function of M2 macrophages.

chemokines to be detected in human adipocytes and were shown to be strongly upregulated following pro-inflammatory stimuli. The primary event in AT inflammation is metabolic dysfunction in adipocytes, followed by the production of cytokines/chemokines, which is then exacerbated by activated ATM, resulting in the recruitment and activation of other immune cells. Various chemokines have been implicated in AT inflammation, among which the most important ones are MCP-1, MCP-2, MCP-3, MCP-4, MIP-1α and MIP-1β. The dominant chemokines in preadipocytes were CCL5, CCL8, CXCL1, and CXCL16, and in adipocytes were CCL6 and CXCL13 [9]. The following chemokines were found in both preadipocytes and adipocytes: CCL2, CCL7,

CCL25, CCL27, CXCL5, CXCL12, and CX3CL1 [9]. Among the various chemokine receptors, CXCR7 was specific for preadipocytes and CXCR2 for adipocytes [9]. These findings indicate the development of a CXCL12-CXCR7 axis which is specific for pre-adipocytes and CXCL5-CXCR2 axis which is specific for adipocytes [9]. In addition to induction of CCL2 and CCL7 in both preadipocytes and adipocytes, EGF specifically enhances CXCL1 and CXCL5 in adipocytes, potentiating the CXCR2-mediated pathway [9]. Visceral adipocytes from insulin-resistant subjects hyperexpressed MCP-1, RANTES, CXCL5/ENA-78, IL-8, lymphotactin-β, and fractalkine. Serum levels of these chemokines are dramatically increased in obesity [10]. The expression of chemokine receptors CCR1, CCR2, CCR3, and CCR5, is elevated in omental and subcutaneous adipose tissues of obese patients. IL-1β was shown to stimulate the secretion of multiple chemokines including MCP-1, IL-8, IP-10, MIP-1α and MCP-4 from mature human adipocytes, with maximum induction noted for IP-10 [11]. Interestingly, hypoxia reduces the expression of chemokines MCP-1 and IL-8 from primary adipocytes. Adiponectin was shown to reduce the secretion of MCP-1, MIP-1α, MIP-1β and IL-8, from adipocytes, in an AMPK-dependent and PPAR-γ- independent pathway. In DIO rats, MCP-1 and XCL-1 was found to be hyperexpressed in subcutaneous and retroperitoneal adipose tissues, XCL-1 was found to be downregulated in epididymal adipose tissue, and SDF-1 showed no significant change in the expression levels in these three different adipose depots.

The interaction between MCP-1 with its receptor CCR2 is considered pivotal in obesity-induced IR in AT. Deletion of MCP-1 or its receptor CCR2 decreased AT inflammation and conferred protection against IR. However, conflicting results were also observed, wherein the loss of MCP-1 neither attenuates obesity-associated macrophage recruitment to WAT nor improves metabolic function [12]. Also, CCR2 deficiency does not normalize ATM content and IR to the levels in lean animals, indicating that MCP-1-CCR2 independent signals also regulate AT inflammation. Overexpression of MCP-1 in the AT, increased macrophage recruitment and induced IR [13]. Among various receptors for MCP-1, CCR2 and CCR5 are the most critical receptors that play a pivotal role in the pathogenesis of IR in AT [14]. CCR2 activation induces the expression of various inflammatory genes and impairs insulin-dependent glucose uptake. The upregulation of MCP-1 and CCL3 from adipocytes may contribute to the development of IR in both adipose and peripheral tissues [15]. Both TLR and NOD1 stimuli are known to inhibit insulin signaling and induce the secretion of cytokines and chemokines. NOD1 was shown to induce the secretion of MCP-1, RANTES, and MIP-2 in mature adipocytes.

Next, to MCP-1, IL-8 is a major adipocytokine produced by adipocytes following stimulation with TNF-α, IL-1β and CRP [16]. IL-8 also induces its own secretion by a positive feedback loop, which is dependent upon MEK-MAPK cascade [16]. Further, IL-8 inhibits insulin-induced Akt phosphorylation and insulin signaling, directly contributing to IR [16]. Circulating levels of IL-8 was found to be significantly increased in DM patients and like other inflammatory markers, IL-8 was found to be hyperexpressed in visceral fat, compared to subcutaneous fat, from insulin resistant subjects. KC (the murine ortholog of human IL-8) expression is increased in the AT and in the plasma of ob/ob and DIO mice. KC expression was seen mainly in the stromal vascular cells and not in adipocytes, and also, the expression is high in pre-adipocytes and decreases with adipocyte maturation. Although KC does not affect adipogenesis, it induces the expression of inflammatory factors and the IR mediators like SOCS3. The lack of KC receptor CXCR2 in hematopoietic cells is sufficient to prevent adipose and skeletal muscle macrophage recruitment and development of IR.

The CXCL12/CXCR4 pathway was recently reported to affect energy metabolism in AT. CXCL12/SDF-1 levels were found to be elevated in DM patients. Mature white adipocytes secrete CXCL12 which induces macrophage infiltration. In addition to this, CXCL12 can directly acts as insulin desensitizing factor in adipocytes thereby worsening insulin sensitivity. In addition to white adipose tissue, CXCL12-CXCR4 pathway plays an essential role in the activation of the brown adipocytes through the P38 and ERK, but not PKA, pathways. Adipocyte specific deletion of CXCR4, in DIO mice, exacerbated obesity. On the contrary, Shin et al. reported that CXCL12 caused IR in white adipose tissues. DIO mice had increased expression of SDF-1 expression in WAT. Treatment of these mice with a CXCR4 antagonist reduced macrophage accumulation and inflammation and improved insulin sensitivity, in AT. CXCL12 plays a dual role in AT. At one end, it recruits M1 macrophages into adipocytes which after differentiation, secretes pro-inflammatory cytokines and induces inflammation leading to IR. On the other hand, it can also have an anti-inflammatory role by mediating T cell polarization toward Tregs and macrophage polarization toward M2 phenotype, both of which are known to impede inflammation [17].

Other chemokines implicated in ATM infiltration are CXCL5 [18] and CXCL14. CXCL5, traditionally regarded as a chemoattractant for neutrophils, is highly expressed in adipose tissue and is mainly expressed in ATMs [18]. It correlates with the fact that plasma levels of CXCL5 were found to be elevated in obese individuals and decrease in these subjects, after a weight reduction program [18]. CXCL5 also blocks insulin signaling by activating the Jak2/STAT5/SOCS2 pathway [18]. CXCL14 exhibits chemoattractive activity for monocytes and dendritic cells. CXCL14 null mice are protected from hyperinsulinemia, obesity-induced IR and hyperglycemia [19]. CXCL14 expression is elevated in white adipose tissue (WAT), in DIO and Ob/Ob mice and the phenotype indicate its involvement in macrophage recruitment and IR [19]. CX3CL1-CX3CR1 pair is an inflammatory adipose chemokine system that modulates monocyte adhesion to adipocytes and is associated with obesity, IR, and DM. Circulating levels of CX3CL1 were found to be elevated in DM and CX3CR1 SNP was associated with central obesity and IR. Studies on CX3CR1 null DIO mice have yielded contrasting results. Polyák et al., have shown reduced macrophage accumulation, attenuated expression of TNF-α, IL-1α and MCP-1 in macrophages, increased expression of lipolytic enzymes and upregulation of thermogenic factors, in the BAT [20]. However, Morris et al. showed that CX3CR1 was not required for the macrophages recruitment into epididymal adipose tissue and CX3CR1 deficiency did not affect IR or hepatic steatosis.

CCR5-mediated signaling in the adipose tissue is also thought to maintain obesity-induced inflammation. As in obese individuals, the expression of CCR5 and its ligands is significantly increased in the WAT of DIO mice [14]. Moreover, a high fat diet causes a stout increase in CCR5+ ATMs in hypertrophic WAT [14]. Also, lack of CCR5 expression in myeloid cells alone was associated with a marked reduction in monocyte infiltration and protects mice from IR [14]. These data suggest that CCR5+ ATMs contribute to the development of obesity-induced adipose tissue inflammation and IR. Kennedy et al. showed that CCR5 has a minor role in regulating macrophage infiltration but increases the influx of CD4+ T cells into hypertrophic AT, indicating that targeting CCR5 may be the best approach to inhibit both macrophage and T cell infiltration.

2.2 Chemokines in skeletal muscle inflammation and insulin resistance

The role of chemokines in skeletal muscle inflammation and insulin resistance is briefly illustrated in **Figure 2**. Compared to AT, skeletal muscle (SM) inflammation

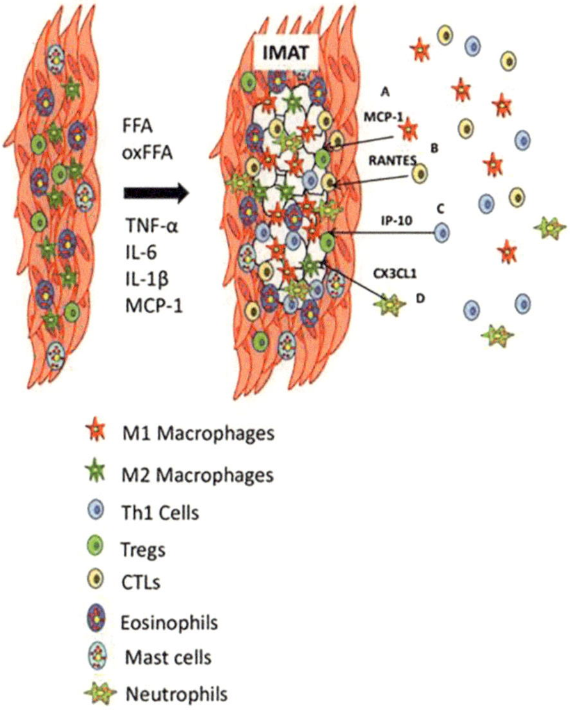

Figure 2.
Chemokines in skeletal muscle inflammation and insulin resistance. During obesity, excess free fatty acids (FFAs) which cannot be stored in the adipose tissue are secreted which undergoes oxidation and gives rise to oxidized free fatty acids (oxFFAs). The FFAs and oxFFAs, along with pro-inflammatory cytokines and chemokines secreted by the AT, set the stage for the formation of intermuscular adipose tissue (IMAT). Transdifferentiation of myocytes into adipocytes and ectopic lipid accumulation into these cells leads to the secretion of cytokines and chemokines. A. MCP-1 attracts M1 macrophages, B. RANTES attracts CTLs, C. IP-10 attracts Th1 cells and D. Fractalkine attracts neutrophils. Eosinophils and mast cell content remains constant.M2 macrophages and Treg content decreases. This sets the stage for SM inflammation and insulin resistance.

and its contribution to IR are less well studied. The main function of insulin in SM is to increase glucose uptake, glycolysis, TCA cycle, ATP synthesis and glycogen synthesis [21]. Under conditions of IR, there is decreased rate of glucose uptake, glycolysis, Krebs cycle, ATP synthesis and glycogen synthesis. Further, the excess circulating FFAs secreted from AT get deposited in SM, leading to the formation of inter-myocellular/inter-muscular AT (IMAT) or perimuscular AT (PMAT) [22]. This ectopic lipid deposition and trans differentiation of myocytes into adipocytes set the stage for inflammation in SM. IMAT refers to fat deposition along the blood vessels, intermuscular space and muscle bundles. Goodpaster published the first study on IMAT and its correlation to IR, in 2000. Histologically, macrophages and T lymphocytes are primarily located on surrounding adipocytes between myocytes, forming the IMAT/PMAT depots. Both are extramyocellular fat that expands substantially in obesity and decreases following weight loss, and both depots are highly correlated with IR. Macrophages and T cells within these adipose depots are markedly increased in obesity [22] and can form crown-like structures surrounding dead or dying adipocytes, as in AT [22]. Additionally, macrophages and T cells can be found at lower frequencies between myofibers in SM [22]. Obesity-linked changes in immune cells and inflammatory markers are much greater in muscle AT than in muscle [22]. Similar to those in visceral AT, immune cells in SM tend to polarize into pro-inflammatory M1 phenotypes in obesity. Most macrophages in SM are CD11c+ and display classical M1 phenotype [22]. Both $CD4^+$ and $CD8^+$ T cells are increased in SM of obese mice [22]. While the proportion of IFN-γ–expressing Th1 cells increases, the proportion of Tregs decreases, in SM [22]. Accordingly, pro-inflammatory markers related to immune cell activation such as TNF-α, IL-1β and IFN-γ are increased [22], while anti-inflammatory markers such as IL-10 are reduced. Both high-fat-high-calorie diet and overfeeding, which are known to induce IR, increased macrophage markers in SM, in healthy subjects. In mice, obesity induced IR was associated with increased accumulation of immune cells, including macrophages and T cells in SM [22]. Like humans, short-term diet induced obese mice have increased macrophage content in SM [22]. Mast cells and eosinophils were also observed in mouse SM but showed no changes with obesity. Changes in other immune cells, including neutrophils, B cells, NK cells, and invariant NKT (iNKT) cells, found in visceral AT, have not been reported in SM during obesity. Similar to adipocytes, SM myocytes express and secrete numerous cytokines such as IL-6, IL-8 and IL-15 and hormones such as FGF-21, irisin, myonectin, and myostatin, collectively called myokines. Whereas most adipokines are pro-inflammatory, regulated by obesity, and involved in the development of obesity-linked metabolic dysfunction, most myokines are anti-inflammatory, secreted during vigorous exercise and counteracts the detrimental effects of adipokines. They have positive effects on glucose and lipid metabolism and dissipate inflammation.

Like adipocytes, skeletal muscle cells also secrete several chemokines, such as MCP-1, MIP-1α, MIP-1β, and IL-8, and express chemokine receptors like CXCR1 and 2 and CCR1, 2, 4, 5, and 10. MCP-1 leads to macrophage infiltration, thereby contributing to the low-grade inflammation. It has been demonstrated that intra-muscular TNF-α expression is restricted to the population of intramuscular leukocytes (mainly, macrophages) and that CCL2 is associated with skeletal muscle inflammation in DIO mice and DM patients [23]. Exposure of myotubes to palmitate resulted in the elevated secretion of CCL2. Muscle-specific overexpression of CCL2 induced macrophage infiltration and altered insulin sensitivity. Moreover, contracting myotubes were shown to secrete CCL2 in a NF-κB manner. Macrophages infiltration is significantly increased in human skeletal muscle from DM patients. TNF-α treated myotubes produce elevated levels of CCL5, CXCL10, CXCL2, IL-8, CCL2, CCL7, CXCL6, CXCL3, and CXCL1. But how many of these chemokines are either directly or indirectly involved in SM inflammation and IR is not known [24].

Recently, CXCL10, which was secreted under the influence of IFN-γ and TNF-α stimulation by myotubes, was shown to recruit Th1 cells, fueling inflammation and IR. In mice, after rigorous exercise, multiplex chemokine analysis showed significant upregulation of CXCL2, CXCL10 and CCL19 and downregulation of CCL5, CCL11, CCL20, CCL21, CXCL1, CXCL9, CXCL10 and CX3CL1, in the SM [25]. Fractalkine (CX3CL1) and its receptor, CX3CR1, were found to play an important role in FFA induced IR in SM cells. Exercise increases the expression of fractalkine in muscle endothelial cells and plays an important role in neutrophil recruitment. Both fractalkine expression and neutrophil recruitment are needed for GLUT4 translocation and secretion of CXCL1 and IL-6 from muscle cells. Further, CX3CL1 is involved in muscle-pancreas cross-talk which would be discussed latter, in this chapter. CXCL1 is another important myokine secreted during palmitate induced lipotoxicity in muscle cells [26]. Overexpression of CXCL1 in skeletal muscle decreased visceral adiposity and increased insulin sensitivity, in DIO mice [27].

2.3 Chemokines in liver inflammation and insulin resistance

The role of chemokines in liver inflammation and insulin resistance is briefly illustrated in **Figure 3**. Next to AT and SM, liver plays a vital role in IR. The liver is the most important metabolic organ which controls glucose and lipid metabolism [28]. The primary function of insulin in the liver is to increase glucose uptake and glycogen synthesis. Further, unlike other insulin target organs, insulin plays a unique role in the liver by inhibiting gluconeogenesis (which occurs only in the liver and to a lesser extent in the kidney). Apart from decreased glucose uptake and glycogenesis (glycogen synthesis), the inability to inhibit gluconeogenesis is the primary cause for high post-prandial blood glucose levels, seen in DM. Unlike SM, ectopic lipid deposition leading to the fatty liver occurs only in morbid obesity and extreme cases of DM. However, chronic inflammation of the liver leading to IR occurs due to the increased circulating levels of pro-inflammatory mediators secreted from AT and to a small

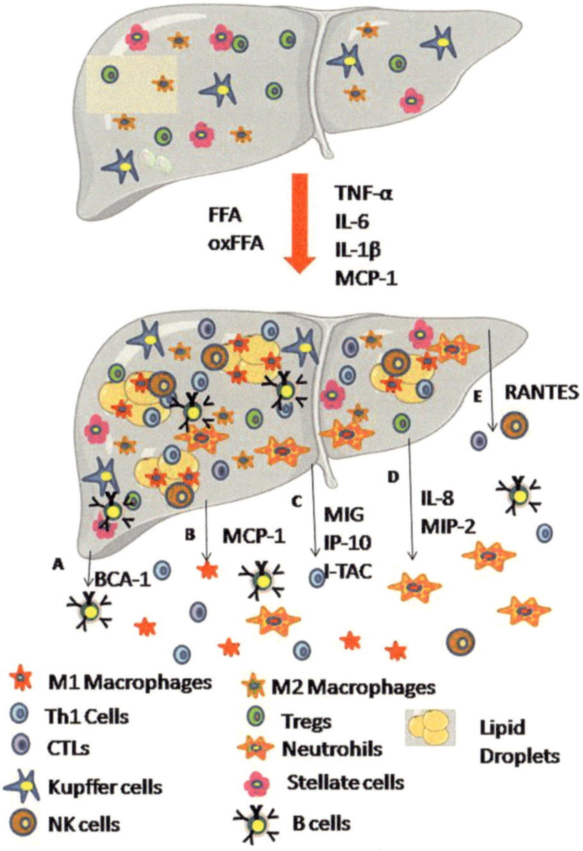

Figure 3.
Chemokines in liver inflammation and insulin resistance. During obesity excess free fatty acids (FFAs) which cannot be stored in the adipose tissue are secreted which undergoes oxidation and gives rise to oxidized free fatty acids (oxFFAs). The FFAs and oxFFAs, along with pro-inflammatory cytokines and chemokines secreted by the AT, set the stage for the formation of ectopic lipid deposition in liver (Hepatic steatosis) Transdifferentiation of myocytes into adipocytes and ectopic lipid. BCA-1 attracts B cells, B. MCP-1 attracts M1 macrophages, C. MIG,IP-10,I-TAC attracts Th1 cells and D. IL-8 and MIP-2 attracts neutrophils and E. RANTES attracts CTLs and NK cells. Kupffer and stellate cell content remains constant but get inflamed. M2 macrophages and Treg content decreases. This sets the stage for liver inflammation and insulin resistance.

extent, from SM. In obesity-induced IR in mice, the resident hepatic macrophages, namely the Kupffer cells, secrete high levels of MCP-1, which recruits circulating macrophages, which in turn augment inflammation and hepatic IR. Kupffer cells are the bona fide liver-resident macrophages and the most abundant cell type in the healthy liver. In DM, activated Kupffer cells become pro-inflammatory and secrete cytokines such as IL-1β, IL-6, TNF-α and MCP-1, which promote inflammation and monocyte recruitment to the liver.

Hepatosteatosis (fatty liver) is described by excessive fat accumulation in the form of triglycerides in the liver. It is generally associated with IR, obesity, DM, dyslipidemia, and metabolic syndromes. Of the many chemokines and their cognate receptors, the hepatic expression of MCP-1-CCR2 is upregulated and plays an important role during the induction of hepatic inflammation and IR.MCP-1 expression is increased in hepatocytes of DIO mice and leads to the hepatic recruitment of CCR2+ myeloid cells which promote hepatosteatosis. Studies show that genetic deletion of MCP-1 and CCR2 attenuates obesity and improves IR and hepatic steatosis. Serum levels of CCL2/MCP-1 were significantly increased in patients with hepatosteatosis with DM.

CCR5 receptor has been recently identified on isolated hepatic stellate cells, which indicates that these hepatic cells are both the source and as well as target for RANTES/CCL5.CCR5 expression is shown to be upregulated in DIO mice, and obese patients [14] and inactivation of CCR5 protects mice from IR and hepatic inflammation [29] and also allows a shift in macrophage polarization toward the M2 phenotype [14]. CCL5 is involved in the recruitment of CCR1+ NK cells and CCR5+ CTLs to the liver in hepatic steatosis. Pharmacological inhibition of CCL5 in mouse models of NAFLD was shown to attenuate liver steatosis, which was likely mediated by the inhibition of lymphocyte chemotaxis. Cenicriviroc, a dual CCR2–CCR5 antagonist, gained much attention owing to its simultaneous effect on two important chemokine pathways, and was shown to be effective in reducing inflammation, steatosis and fibrosis in the liver [30].

T cell chemokines like CXCL9 (MIG), CXCL10 (IP-10) and CXCL11 (MIP-2) have been shown to promote cholesterol-induced steatohepatitis and facilitate lipid accumulation. The pharmacological blockade of CXCR3 in mice was shown to prevent the Development and also reverse established steatohepatitis. Serum levels of CXCL9 are associated with liver fibrosis in patients [31] and CXCL10 has been identified as an independent risk factor for NASH. CXCL10 not only induces inflammation but also directly promotes steatosis by stimulating lipogenesis and promoting macrophage-associated liver injury in mouse models of NASH [32]. B cell recruitment to the liver is mediated by the upregulation of CXCL13 (BCA-1) [33].

Hepatosteatosis induces neutrophil migration by upregulating CXCL1 and CXCL8 in hepatocytes. Treatment of postpartum cows with recombinant IL-8 increased milk production but also increased hepatosteatosis. G31P, an antagonist of CXCL8, improved hepatic insulin sensitivity by modulating the expression of genes related to gluconeogenesis, in db/db mice [34]. The expression of several genes involved in de novo lipogenesis were decreased in treated mice [34]. Immune cell infiltration and cytokine secretion were also attenuated in these mice [34]. G31P also improved the ratio of M1 and M2 macrophages [34]. Furthermore, G31P ameliorates metabolic disturbances by inhibiting CXCR1 and CXCR2 pathways in these mice [34].

2.4 Chemokines in beta cell inflammation and loss

The role of chemokines in beta cell inflammation and loss is illustrated in **Figure 4**. Finally, chronic DM commonly presents itself with decreased β-cell function, often referred to as loss of β-cell mass. Under clinical condition, frank DM sets in only after the pancreatic β cells stop producing insulin. Numerous efforts have been made, to elucidate the mechanisms behind β-cell loss in DM. Oxidative stress, endoplasmic reticulum (ER) stress, hypoxia stress, inflammation and protein aggregation are all involved in the β-cell loss. In response to these stressors, β cells can either undergo apoptosis or uncontrolled autophagy [35]. Emerging evidence also suggests that they can dedifferentiate or transdifferentiate into other pancreatic cell types, which is a recent fast emerging concept, in the pathogenesis of β-cell dysfunction.

In DM, the primary mechanism leading to the decreased β-cell mass is apoptosis. While the toxic role of pro-inflammatory cytokines in inducing β -cell death is well documented, the role played by chemokines in this process is less well studied. CXCL10 (IP-10), was the first chemokine expressed in the pancreas in a mice model

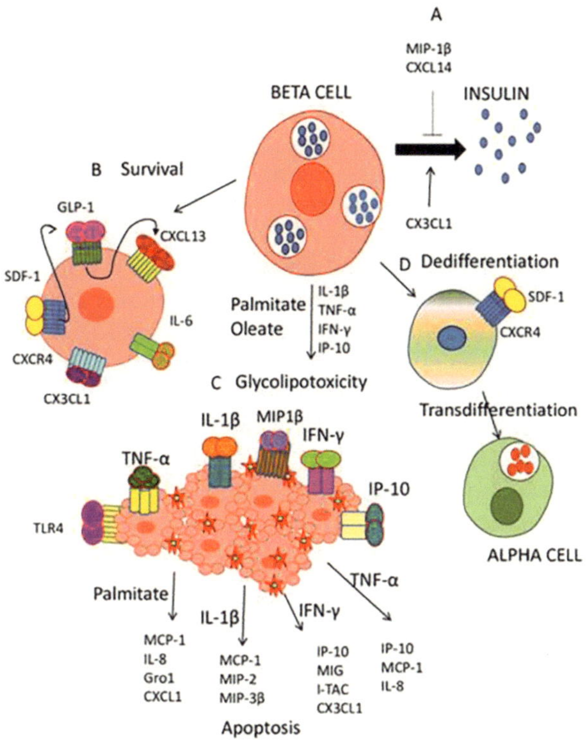

Figure 4.
Chemokines in beta cell inflammation and loss. During obesity, excess free fatty acids (FFAs) which cannot be stored in the adipose tissue are secreted which undergoes oxidation and gives rise to oxidized free fatty acids (oxFFAs). The FFAs and oxFFAs, along with, pro-inflammatory cytokines secreted by the AT and hyperglycemia sets the stage for beta cell inflammation. A. Chemokines like MIP-1β and CXCL14 can directly inhibit insulin secretion while CX3CL1 can augment the secretion. B. Chemokines like SDF-1, CXCL13 and CX3CL1 can counteract the toxic effect of pro-inflammatory cytokines and can increase the survival of beta cells along with survival factors like IL-6 and GLP-1. C. Glycolipotoxicity leads to ectopic lipid deposition and secretion of macrophage attractant MCP-1. The recruited macrophages secrete pro-inflammatory cytokines TNF-α and IL-1β, which along with palmitate activated TLR4, induce beta cell apoptosis. IFN-γ secreted by beta cells augment this effect. These cytokines also induce the secretion of plethora of chemokines out of which some like IP-10 and MIP-1β can have direct cytotoxic effect. D SDF-1 can promote dedifferentiation of beta cells and can thereby protect them from glycolipotoxicity. Under certain conditions these cells transdifferentiate into glucagon producing alpha cells.

of LCMV-induced DM. Islets isolated from patients with DM secreted high levels of CXCL10. Treatment of human islets with CXCL10 decreased cell viability, impaired insulin secretion, and reduced insulin expression, through PI3K dependent signaling [36]. These effects are found to be independent of its receptor CXCR3 and were mediated through its interaction with TLR4 [36]. Overexpression of CXCL10 in isolated mouse islets leads to enhanced lymphocyte infiltration and increased apoptosis. CXCL10 antagonist NBI-74330 was shown to significantly reduce β-cell loss in a streptozotocin induced DM rat model.

Along with β-cell destruction, de-differentiation and re-differentiation of β-cell into α-cells occurs in DM. CXCL12 (SDF-1) regulates the differentiation and function of immune cells and also play an anti-inflammatory and immunomodulatory role. Elevated levels of SDF-1 are commonly seen in DM subjects [37] and are associated with diabetic insulitis, nephropathy and adipose tissue inflammation. DPP-IV (Dipeptidyl peptidase IV) inhibitor saxagliptin improved the function of β-cells by regulating SDF-1 expression. Several studies have revealed the importance of SDF-1 in β cell survival, after islet transplantation [38]. Furthermore, during the terminal differentiation of β cells, SDF-1 prevents apoptosis by activating the PI3K/AKT and WNT/β-catenin pathways [38]. SDF-1 can bind to CXCR4 and upregulate FOXO1 expression, thereby inhibiting the dedifferentiation of pancreatic β cells. Hyperglycemia causes a partial loss of SDF-1 activity, which is then unable to bind to CXCR4 and inhibit the dedifferentiation of pancreatic β cells.

Bone marrow derived mesenchymal stem cells express a restricted set of chemokine receptors (CXCR4, CX3CR1, CXCR6, CCR1 and CCR7) and, accordingly, show migration in response to the chemokines like CXCL12, CX3CL1, CXCL16, CCL3, and CCL19. Migration of these cells into pancreatic islets and their differentiation into β cells was largely mediated through CX3CL1 and CXCL12 chemokines. The muscle-pancreas intercommunication axis, which is largely responsible for improved β-cell function, in DM patients, routinely undergoing rigorous physical exercise, is largely mediated by CX3CL1 [39]. CX3CL1 decreases glucagon secretion and protects β-cells from TNF-α induced apoptosis [39]. CX3CR1 knockout mice have impaired glucose tolerance, resulting from decreased insulin secretion. CX3CL1 administration improved glucose tolerance and induced insulin secretion, in DIO mice.

3. Chemokines in tuberculosis

3.1 Chemokines in granuloma formation

The role of chemokines in granuloma formation is briefly illustrated in **Figure 5**. Granulomatous response is a very characteristic feature of TB infection. TB granuloma is defined as a focal aggregate of immune cells that forms in response to TB bacilli invasion [40]. A basic requirement of granuloma formation involves directed migration and organization of immune cells, so as to curtail the infection. This directorial migration and organization of cells is facilitated mainly through the orchestration of cytokines, chemokines and cell adhesion molecules. Chemokines apart from recruitment of immune cells into the granuloma, are also involved in arresting the emigration of cells out of the granuloma. If the activated immune cells are not controlled tightly within the granuloma, they might enter circulation and aggravate systemic inflammation. To restate, one of the crucial roles of granuloma is not only to localize and contain the bacteria but also to shield the inflammatory

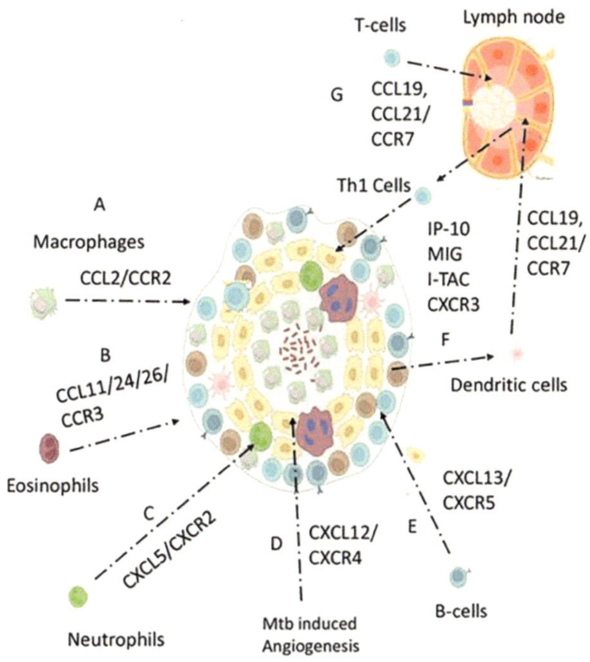

Figure 5.
Chemokines in TB granuloma formation and maintenance. A. The most important event in the formation of TB granuloma is the interaction between M.tb and lung macrophages. This leads to the secretion of MCP-1 which attracts CCR2+ macrophages. Thus, during early infection CCL2/CCR2 axis leads to macrophage recruitment and infection B. Exotoxins (-1, -2 and -3) secreted by the granuloma attract CCR3+ eosinophils. C. CXCL5 secreted by the granuloma attracts CXCR2+ neutrophils. D. the SDF-1/CXCR4 axis plays an important role in the recruitment of endothelial cells and angiogenesis. E. Peripheral B cells are recruited viz CXCL13/CXCR5 axis. F. The immature DCs in lung undergo maturation upon Mtb infection and navigate to the draining lymph node through CCL19/CCL21/CCR7 axis. G. Interestingly, the same axis guides the naive T cells into the lymph nodes were T cell-DC interaction and priming takes place. Finally, Mtb specific activated Th1 cells get recruited into the granuloma through a panel of IFN-γ induced chemokines (IP-10, MIG, I-TAC) and their cognate receptor (CXCR3). Containment of Tb growth inside the granuloma takes place only after the timely recruitment of fully activated Th1 cells.

response from spilling over to systemic circulation. A delicate balance between pro- and anti-inflammatory cytokines plays a major role in the formation and maintenance of TB granulomas: TNF-α, IL-1β and IFN-γ are particularly important in promoting the formation and function of the granuloma, whereas IL-10 and IL-1Ra are necessary to control the inflammatory response, within the granuloma.

Caseous granulomas are typical of TB. These structures are formed by epithelioid macrophages surrounding a cellular necrotic region with a rim of lymphocytes (T- and B-cells). Numerous chemokine families are involved in granuloma formation. Some chemokines are produced by respiratory epithelium and others are produced by immune cells themselves. In particular, CCL2, CCL12 and CCL13 are important for the early recruitment of macrophages. Osteopontin, a highly phosphorylated sialoprotein present in mineralized extracellular matrix of bones and teeth, produced by macrophages and lymphocytes, promotes the adhesion and recruitment of these cells. CCL19 and perhaps CCL21 are involved in the recruitment and priming of IFN-γ-producing T cells [41]. Both CCL19 and CCL21 bind CCR7 and direct the homing of naive, central memory T cells and DCs to secondary lymphoid organs. Studies pointed out that CXCL13 is not involved in generating IFN-γ responses, but is essential for the spatial arrangement of lymphocytes within the granuloma and for optimal activation of macrophages, and subsequent control of mycobacterial growth [41]. CXCL13 is also involved in B-cell recruitment and the formation of follicular structures [41].

CCL5(RANTES) is a Th1 chemokine which promotes T cell activation and proliferation and aids in chemotaxis of memory T cells to the granuloma. Several studies strongly connect the link between CCL5 and disease susceptibility, due to their peculiar role in regulation of T-cell activation and expression during *M.tb* infection. Lymphoid structure organization and inducing protective immune response during *M.tb* infection are dependent on homeostatic chemokine expression. Specifically, CCL19/CCL21 are required for optimal priming, activation, and agglomeration of T cells in lungs, likely through their role in migration of DCs and priming of Ag-specific T cell responses in lymphoid organs [41]. Initiation and organization of lymphoid areas within granuloma of infection is significantly disrupted due to the absence of homeostatic chemokines like CCL19 and 21.

In accordance to changes in chemokine profile, differential expression of chemokine receptors occurs in immune cells, as part of cellular maturation/activation [42]. For instance, CXCR1 and CXCR2 are highly expressed by neutrophils permitting responses to chemokine ligands CXCL1–8 [42]. On the contrary, CXCR3 is expressed by effector lymphocytes and responds to CXCL9,10 and 11 [42]. Animal studies involving different antigens induced granulomas showed that PPD-elicited granulomas alone had higher expression of neutrophil chemotactins like CXCL2, CXCL5 and CCL3, which correlates with increased neutrophil influx, during early stages of granulomas. Also, during adaptive granuloma formation, the CXCL chemokines (CXCL2, CXCL5, CXCL9, CXCL10, CXCL11) increased up to 10-fold over the innate response [42]. Polymorphonuclear Neutrophils (PMN) are necessary for early granuloma formation during chronic *M.tb* infection without influencing mycobacterial growth restriction [43]. PMN mediated regulation of granuloma formation depends on chemokine signaling through CXCR3, in particular MIG [43]. Absence of CXCR3 did not show any defects in bacterial control; however, granuloma formation was impaired with associated decrease in the number, size, and density of granulomas in the lungs [43]. Another report shows that CXCR4 plays a critical role in the granuloma formation by sustaining angiogenesis [44].CCR2 is the logical candidate to recruit monocyte/macrophages to the site of infection as macrophages and monocytes are chemotactic toward CCR2 ligands, CCL2, CCL7, and CCL12 [45]. The absence of CCR2 resulted in prolonged macrophage deficiency as well as delayed T-cell migration, particularly CD4 T cells, during early stages of infection, in the lungs [45]. Delayed T cell migration leads to significantly delayed IFN-γ response and impaired containment of *M.tb* [45].

3.2 Chemokines in latent tuberculosis

The WHO defines LTBI as a state of persistent immune response to stimulation by *M.tb* antigens without evidence of clinically manifested active TB. According to recent estimates, approximately one-quarter of the global population is infected with LTBI [46]. The duration of latency is variable, and healthy individuals can harbor LTBI for their entire lifetime. In a small fraction (~5–15%) of individuals, reactivation occurs, often within the first 2–5 years following infection. Reactivation is defined as a process by which a dormant subclinical latent infection transforms into active TB disease. Hence, individuals with LTBI represent a major reservoir for new active TB cases.

Chemokines are known to play a key role in establishing LTB infection. CCL1, CRP, CXCL10, and vascular endothelial growth factor (VEGF) were found to be strongest differentiating markers between active and Latent TB(LTB) [47]. Increased concentrations of these inflammatory mediators reflect the activity of infected macrophages, innate immune cells such as NK cells, innate lymphoid cells (ILCs) as well as activated T lymphocytes. VEGF and CXCL10 produced by APCs regulate cell growth and chemotaxis and may act as driving forces for stimulated angiogenesis observed in ATB lesions. Individuals with latent TB were shown to have increased CXCR2 expression. Altered CCL3 and CCL4 levels could potentially modify T cell recruitment in LTBI. A study also found a significant positive correlation between BMI and various chemokine levels including CCL1, CCL4,CCL11,CXCL2,CXCL9 and CXCL11 levels in LTBI individuals suggesting that LTBI subjects with low BMI has diminished levels of various chemokines posing an increased risk of developing active TB [48].

IFN-γ is the classical marker for detecting LTB. Besides IFN-γ, there may be several cytokines and chemokines that have been investigated as potential biomarkers for *M.tb* infection and disease. For instance, IL-4,IL-6,IL-10,CCL8,CXCL8 and CXCL10 are closely linked to active TB. Recent studies suggest that IP-10 along with the combination of IFN-γ may enhance diagnostic performance of IGRA to detect LTB, especially in young children [49]. Moreover, studies show that IP-10 as an individual biomarker can be improved when combined with IL-7 to differentiate active TB and healthy controls. Also, when IP-10 combines with BCA-1 can differentiate active TB and LTBI. A study conducted by Liu et al., showed elevated levels of IL-2, IP-10, CXCL11 and CXCL12 in patients with TB and in a sub-group participant with LTBI who have showed a higher level of IFN-γ producing cells by ELISPOT assay compared with other latently infected individuals. Monocytes isolated from ATB and LTB patients express CXCR1 and CXCR2 [50]. After stimulation with purified protein derivatives (PPD), the in vitro levels of CXCL8 were below the median levels of all patients with prior TB.

ATB individuals exhibited significantly higher levels of CCL1, CCL3, CXCL1, CXCL2, CXCL9 and CXCL10 compared to LTB and control individuals. ATB patients with bilateral or cavitary disease displayed significantly elevated levels of CCL1, CCL3, CXCL1, CXCL10 and CXCL11 compared to patients with unilateral or non-cavitary disease and also revealed a significant positive relationship with bacterial burdens. In addition, PTB individuals with delayed culture conversion displayed elevated levels of CCL1, CCL3, CXCL1 and CXCL9 at the time of diagnosis. The chemokine levels were significantly reduced following successful anti-TB treatment. Thus, ATB was associated with elevated levels of chemokines, which are partially reversed following therapy. Chemokines also serve as serum biomarkers of disease severity, predicting bacterial burden and delayed culture conversion. In another study conducted by the same group, CCL2, CCL3, CCL4, CXCL8, CXCL10, and CX3CL1 levels were found to be increased, while CXCL1 levels were decreased in ATB patients, compared to control. Similarly, elevated levels of CCL3,CXCL8 and CXCL10 were strongly associated with increased risk of unfavorable treatment.

3.3 Chemokines in immunity against TB

Chemokines being a key player in recruitment of leukocytes to a site of infection, plays a dynamic role in TB containment. The availability of various animal models of TB over the past decade has made it possible to decipher several key mechanisms driven by chemokines that mediate anti TB immunity. *M.tb*, upon access into the lung, is taken up by alveolar macrophages, where *M.tb* multiplies while evading the macrophage killing mechanism. In addition to this, infected macrophages secrete chemokines and cytokines, which results in the recruitment and activation of various immune cell populations to the lung. Around 12 days post-infection in low dose aerosol infected mouse model, there is an early influx of various innate immune cells into lungs including NK cells, dendritic cells, neutrophils, γδ T cells and monocyte-derived macrophages. Chemokines orchestrate the specific recruitment of these innate immune cells to the lungs. Especially, increased expression of CXCL3 and CXCL5 is observed as early as day 12 after infection and correlates with early influx of neutrophils and NK cells, which express CXCR2, the cognate receptors for these chemokines. Lung epithelial cells can directly sense *M.tb* through TLR2 present throughout the human airway epithelium and can promote chemokine secretion, resulting in immune cell recruitment. TLR2 promote the innate immune system to recognize numerous "pathogen associated molecular patterns" present on the *M.tb* structure. In response to *M.tb* stimulation, CCL2 and CXCL8 are produced by alveolar and bronchial epithelial cells [51]. In addition to this, CXCL5 is also shown to be secreted by lung epithelium which signals through CXCR2 can increase neutrophil influx [52]. Even with the accumulation of these innate immune cells, *M.tb* continues to grow exponentially over the first 2–3 weeks following infection. At this stage, the adaptive immune component gets activated and effector T cells get recruited to the site of infection.

Lung resident DCs can take up live *M.tb* within the lungs and transport them to the lungs draining mediastinal lymph nodes, where they were thought to serve as APCs. Migration of DCs is governed by chemokine-receptor interactions, and this occurs around day 14 post infection in the mouse model of TB. Uptake of *M.tb* by DCs leads to the upregulation of CCR7, which guides the cells to the mediastinal lymph node following a gradient of the homeostatic chemokines CCL19 and CCL21. CCL21 directs the initial migration of DCs. Notably, DCs from mice lacking CCR7 have an impaired ability to migrate to the draining lymph nodes, resulting in delayed priming of *M.tb*-specific T cells [53]. Recently, it has come to light that the cell populations that become infected and carry antigen to the lymph node, and those that directly prime the T cells, are different. Indeed, infected CCR2+ inflammatory monocytes are important for antigen delivery into the lung, where they release soluble antigen that can be taken up and presented by resident lymph node DCs. Subsequent recognition of *M.tb* antigens by naïve T cells bearing specific T cell receptors, in the presence of co-stimulatory signals and adequate cytokines, in the microenvironment leads to the activation, proliferation and differentiation of naïve T cells into effector cells.

Induction of inflammatory chemokines during *M.tb* replication ultimately results in the recruitment of activated effector T cells, from the periphery. The activated T cells which exit lymph nodes are now able to enter lungs via circulation through ligation of surface endothelial receptors that are upregulated in response to inflammation. Several chemokines and their associated receptors have been linked to T cell migration into the lung, during TB infection. Upon commitment to the Th1 subset, the main CD4+ T cell subset implicated in *M.tb* control, namely the Th1 effector T cells upregulate the chemokine receptors CXCR3 and CCR5. It is believed that this is directly related to their recruitment into the infected lung, as the ligands for these receptors, CXCL9-11 for CXCR3 and CCL3-5 and -8 for CCR5, are upregulated in the lungs of *M.tb*-infected mice. Several mechanistic studies have addressed the requirement for CXCR3 and CCR5 expression on T cells, providing evidence that there is significant redundancy in the expression of these inflammatory chemokines and their receptors on *M.tb*-specific T cell recruitment to the lung [43, 54]. Human studies have also shown associations between mutations in CCL2 and CCL5 with pulmonary TB.

Upon entry into the lung parenchyma, however, proper *M.tb* containment is dependent upon the proper juxtaposition of effector T cells with *M.tb*-infected macrophages. In recent times, several reports have demonstrated the expression of several homeostatic chemokine, which is commonly seen in secondary lymphoid organs(SLO) in *M.tb* infected lungs [41]. Such chemokines, including CCL19, CCL21, CXCL12 and CXCL13, drive the organization of lymphoid follicles in secondary lymphoid organs(SLOs) in the periphery [41]. Ectopic lymphoid follicles consisting of stromal and lymphoid aggregates, have been reported in chronic infection and inflammation. Interestingly, during *M.tb* infection in mice, non-human primates and humans $CD4^+$ T cells expressing CXCR5 receptors accumulate in the lungs, within ectopic lymphoid follicles [55]. Strikingly, these $CD4^+$ cells bearing CXCR5 receptors produce high levels of pro-inflammatory cytokines and upon accumulation in the lung, respond to CXCL13 likely produced by stromal cells early during infection, and localize near *M.tb*-infected macrophages to mediate *M.tb* control [55]. Consistently, both CXCR5 and CXCL13 deficient mice lacked the formation of ectopic lymphoid follicles and displayed decreased control of *M.tb* thus indicating the non-redundant role for CXCR5-CXCL13 axis in TB [55]. CXCR5 deficiency resulted in localization of $CD4^+$ T cells around blood vessels in the *M.tb*-infected lungs, forming perivascular cuffs indicative of their inability to localize in opposition to infected macrophages [55]. So, not only the timely induction of chemokine mediated T cell recruitment to the lung is critical for *M.tb* control but also the chemokines play a critical role in positioning the *M.tb* specific T cells with *M.tb* infected macrophages within the lung parenchyma for effective *M.tb* control. Indeed, early vaccine-induced production of CXCL9, CXCL10 and associated recruitment of CXCR3-expressing T cells is beneficial in vaccine-induced protection against *M.tb* challenge. In addition, vaccine strategies that induce early CXCL13 production to enhance and improve early T cell localization near *M.tb*-infected macrophages can be harnessed for vaccine design against TB [56]. Together, there is hoarding evidence to show that chemokines induced in response to *M.tb* infection mediate DC trafficking to lymph nodes, recruitment of activated T cells to the site of the lungs, effective localization of T cells within the lung parenchyma and juxtaposition of *M.tb* specific T cells and *M.tb* infected macrophages to facilitate *M.tb* containment. Though all these processes depend upon chemokines, they often do not completely eliminate the bacteria. Further understanding of the mechanisms that lead to *M.tb* containment will allow better development of novel therapies against TB.

4. Chemokines in diabetes-tuberculosis synergy

Compared to studies looking at the role of chemokines in DM and TB, the number of studies, looking at chemokine levels in DM-TB co-morbidity, is few. In a recent study, the levels of IP-10, IL-8 and SDF-1 were quantified in TB patients with various grades of glucose intolerance [57]. IP-10 levels were significantly reduced in TB patients across various groups of glucose intolerance [57]. Circulating levels of IL-8 levels were found to be reduced in DM with increasing grades of glucose intolerance [57]. ATB decreased the IL-8 levels in newly diagnosed DM subjects, and increased levels in chronic DM subjects, who were under treatment [57]. ATB increased the levels of SDF-1 in control and pre-diabetic subjects [57]. Thus, in ATB, based on glucose intolerance, chemokines levels showed drastic fluctuations, which indicate compromised immunity. In the highly sensitive guinea pig model of TB, infected animals had severe and rapidly progressive TB with decreased survival rate, more severe pulmonary and extra-pulmonary pathology, and a higher bacterial burden in DM animals compared to control animals [58]. These animals had an exacerbated pro-inflammatory response with more severe granulocytic response with hyperexpression of several pro and anti-inflammatory cytokines along with IL-8 in the lungs and IL-8 and MCP-1 in the spleen [58]. TB disease progression was identical in both groups during the early stages but was more severe by day 90 in the diabetic pigs [58]. When db/db mice were infected with *M.tb*, a markedly increased bacterial load in their lungs was seen, compared to wild-type mice [59]. They also had highly disorganized granulomas, neutrophilia, and reduced B cells in the lungs, correlating with dysregulated expression of XCL1, CCL2, CXCL1, CXCL2, and CXCL13 [59]. Although the Th1 cell response developed normally, production of pulmonary IFN-γ was delayed and was ineffective [59]. Monocytes from DM subjects when infected with *M.tb* produced significantly reduced levels of IL-8 compared to control subjects, indicating a functional defect in these monocytes in combating *M.tb* infection [60].

5. Conclusion

To conclude chemokines and chemokine receptors play an important role in both DM, as well as TB. In DM, as was seen in this chapter, they play an important role both in IR and pancreatic β cell loss. In the adipose tissue, at least two different immune circuits can be deciphered- one associated with homeostasis and insulin sensitivity and another associated with inflammation and IR. The immune cells, cytokines and chemokines involved in these two circuits are completely different. In the skeletal muscles chemokines play an important role in immune cell recruitment and formation of intramuscular adipose tissue which impairs insulin sensitivity. In the liver, chemokines were directly involved in steatohepatitis and IR. In pancreatic β cells, chemokines were involved both in augmenting and inhibiting insulin secretion. They also play a decisive role in survival versus apoptosis. Interestingly, some are also involved in dedifferentiation and redifferentiation of β cells into α cells. With respect to TB, chemokines are involved in the formation and maintenance of granulomas, latent infection and immunity against active TB. Recent, advancement in chemokine research has led to the discovery of several antagonists and agonists to chemokine receptors, out of which some have entered clinical trials. In immediate future, we feel, at least some of them might find their use as drugs which can break the DM-TB synergy.

Acknowledgements

This work was supported by Department of Biotechnology, New Delhi (BT/PR20621/MED/29/1207/2017).

Conflict of interest

None.

Author details

Vivekanandhan Aravindhan* and Srinivasan Yuvaraj
Department of Genetics, Dr. ALM PG IBMS, University of Madras, Chennai, India

*Address all correspondence to: cvaravindhan@gmail.com

References

[1] World Health Organization. Global tuberculosis report 2016. World Health Organization. 2016. Available from: https://apps.who.int/iris/handle/10665/2504414

[2] WHO. Tuberculosis: A global Emergency. World health 1993;**46**(4): 3-31. Available from: https://apps.who.int/iris/handle/10665/52639

[3] International Diabetes Federation. IDF Diabetes Atlas, 10th ed. Brussels, Belgium; 2021. Available from: https://www.diabetesatlas.org

[4] Yorke E, Atiase Y, Akpalu J, Sarfo-Kantanka O, Boima V, Dey ID. The bidirectional relationship between tuberculosis and diabetes. Tuberculosis Research and Treatment. 2017;**2017**:1702578

[5] Domingo-Gonzalez R, Prince O, Cooper A, Khader SA. Cytokines and chemokines in mycobacterium tuberculosis infection. Microbiology Spectrum. 2016;**4**(4):5-23

[6] Aravindhan V, Madhumitha H. Metainflammation in diabetic coronary artery disease: Emerging role of innate and adaptive immune responses. Journal Diabetes Research. 2016;**2016**:6264149

[7] Chen K, Bao Z, Tang P, Gong W, Yoshimura T, Wang JM. Chemokines in homeostasis and diseases. Cellular & Molecular Immunology. 2018;**15**:324-334

[8] Gastaldelli A, Gaggini M, DeFronzo RA. Role of adipose tissue insulin resistance in the natural history of Type 2 diabetes: Results from the San Antonio metabolism study. Diabetes. 2017;**66**:815-822

[9] Kabir SM, Lee ES, Son DS. Chemokine network during adipogenesis in 3T3-L1 cells: Differential response between growth and proinflammatory factor in preadipocytes vs. adipocytes. Adipocytes. 2014;**3**:97-106

[10] Huber J, Kiefer FW, Zeyda M, Ludvik B, Silberhumer GR, Prager G, et al. CC chemokine and CC chemokine receptor profiles in visceral and subcutaneous adipose tissue are altered in human obesity. The Journal of Clinical Endocrinology and Metabolism. 2008;**93**:3215-3221

[11] Alomar SY, Zaibi MS, Kepczynska MA, Gentili A, Alkhuriji A, Mansour L, et al. PCR array and protein array studies demonstrate that IL-1beta (interleukin-1beta) stimulates the expression and secretion of multiple cytokines and chemokines in human adipocytes. Archives of Physiology and Biochemistry. 2015;**121**:187-193

[12] Kirk EA, Sagawa ZK, McDonald TO, O'Brien KD, Heinecke JW. Monocyte chemoattractant protein deficiency fails to restrain macrophage infiltration into adipose tissue. Diabetes. 2008;**57**:1254-1261

[13] Kamei N, Tobe K, Suzuki R, Ohsugi M, Watanabe T, Kubota N, et al. Overexpression of monocyte chemoattractant protein-1 in adipose tissues causes macrophage recruitment and insulin resistance. The Journal of Biological Chemistry. 2006;**281**:26602-26614

[14] Kitade H, Sawamoto K, Nagashimada M, Inoue H, Yamamoto Y, Sai Y, et al. CCR5 plays a critical role in obesity-induced adipose tissue inflammation and insulin resistance

by regulating both macrophage recruitment and M1/M2 status. Diabetes. 2012;**61**:1680-1690

[15] Gerhardt CC, Romero IA, Cancello R, Camoin L, Strosberg AD. Chemokines control fat accumulation and leptin secretion by cultured human adipocytes. Molecular and Cellular Endocrinology. 2001;**175**:81-92

[16] Kobashi C, Asamizu S, Ishiki M, Iwata M, Usui I, Yamazaki K, et al. Inhibitory effect of IL-8 on insulin action in human adipocytes via MAP kinase pathway. Journal of Inflammation Research(Lond). 2009;**6**:25

[17] Karin N. The multiple faces of CXCL12 (SDF-1alpha) in the regulation of immunity during health and disease. Journal of Leukocyte Biology. 2010;**88**:463-473

[18] Chavey C, Lazennec G, Lagarrigue S, Clape C, Iankova I, Teyssier J, et al. CXC ligand 5 is an adipose-tissue derived factor that links obesity to insulin resistance. Cell Metabolism. 2009;**9**:339-349

[19] Hara T, Nakayama Y. CXCL14 and insulin action. Vitamins and Hormones. 2009;**80**:107-123

[20] Polyak A, Winkler Z, Kuti D, Ferenczi S, Kovacs KJ. Brown adipose tissue in obesity: Fractalkine-receptor dependent immune cell recruitment affects metabolic-related gene expression. Biochimica et Biophysica Acta. 2016;**1861**:1614-1622

[21] Sylow L, Tokarz VL, Richter EA, Klip A. The many actions of insulin in skeletal muscle, the paramount tissue determining glycemia. Cell Metabolism. 2021;**33**:758-780

[22] Khan IM, Perrard XY, Brunner G, Lui H, Sparks LM, Smith SR, et al. Intermuscular and perimuscular fat expansion in obesity correlates with skeletal muscle T cell and macrophage infiltration and insulin resistance. International Journal of Obesity. 2015;**39**:1607-1618

[23] Patsouris D, Cao JJ, Vial G, Bravard A, Lefai E, Durand A, et al. Insulin resistance is associated with MCP1-mediated macrophage accumulation in skeletal muscle in mice and humans. PLoS One. 2014;**9**:e110653

[24] Tourniaire F, Romier-Crouzet B, Lee JH, Marcotorchino J, Gouranton E, Salles J, et al. Chemokine expression in inflamed adipose tissue is mainly mediated by NF-kappaB. PLoS One. 2013;**8**:e66515

[25] Little HC, Tan SY, Cali FM, Rodriguez S, Lei X, Wolfe A, et al. Multiplex quantification identifies novel exercise-regulated myokines/cytokines in plasma and in glycolytic and oxidative skeletal muscle. Molecular & Cellular Proteomics. 2018;**17**:1546-1563

[26] Masuda S, Tanaka M, Inoue T, Ohue-Kitano R, Yamakage H, Muranaka K, et al. Chemokine (C-X-C motif) ligand 1 is a myokine induced by palmitate and is required for myogenesis in mouse satellite cells. Acta Physiologica. Oxford, England. 2018;**222**:e12975

[27] Pedersen L, Olsen CH, Pedersen BK, Hojman P. Muscle-derived expression of the chemokine CXCL1 attenuates diet-induced obesity and improves fatty acid oxidation in the muscle. American Journal of Physiology, Endocrinology and Metabolism. 2012;**302**:E831-E840

[28] Rui L. Energy metabolism in the liver. Comprehensive Physiology. 2014;**4**:177-197

[29] Xu L, Kitade H, Ni Y, Ota T. Roles of chemokines and chemokine receptors

in obesity-associated insulin resistance and nonalcoholic fatty liver disease. Biomolecules. 2015;**5**:1563-1579

[30] Lefebvre E, Moyle G, Reshef R, Richman LP, Thompson M, Hong F, et al. Antifibrotic effects of the dual CCR2/CCR5 antagonist cenicriviroc in animal models of liver and kidney fibrosis. PLoS One. 2016;**11**:e0158156

[31] Wasmuth HE, Lammert F, Zaldivar MM, Weiskirchen R, Hellerbrand C, Scholten D, et al. Antifibrotic effects of CXCL9 and its receptor CXCR3 in livers of mice and humans. Gastroenterology. 2009;**137**(309-19):19 e1-3

[32] Tomita K, Freeman BL, Bronk SF, LeBrasseur NK, White TA, Hirsova P, et al. CXCL10-mediates macrophage, but not other innate immune cells-associated inflammation in murine nonalcoholic steatohepatitis. Scientific Reports. 2016;**6**:28786

[33] Bigorgne AE, Bouchet-Delbos L, Naveau S, Dagher I, Prevot S, Durand-Gasselin I, et al. Obesity-induced lymphocyte hyperresponsiveness to chemokines: A new mechanism of Fatty liver inflammation in obese mice. Gastroenterology. 2008;**134**:1459-1469

[34] Cui S, Qiao L, Yu S, Men L, Li Y, Li F, et al. The antagonist of CXCR1 and CXCR2 protects db/db mice from metabolic diseases through modulating inflammation. American Journal of Physiology. Endocrinology and Metabolism. 2019;**317**:E1205-E1E17

[35] Cerf ME. Beta cell dysfunction and insulin resistance. Frontiers in Endocrinology (Lausanne). 2013;**4**:37

[36] Schulthess FT, Paroni F, Sauter NS, Shu L, Ribaux P, Haataja L, et al. CXCL10 impairs beta cell function and viability in diabetes through TLR4 signaling. Cell Metabolism. 2009;**9**:125-139

[37] Derakhshan R, Arababadi MK, Ahmadi Z, Karimabad MN, Salehabadi VA, Abedinzadeh M, et al. Increased circulating levels of SDF-1 (CXCL12) in type 2 diabetic patients are correlated to disease state but are unrelated to polymorphism of the SDF-1beta gene in the Iranian population. Inflammation. 2012;**35**:900-904

[38] Yano T, Liu Z, Donovan J, Thomas MK, Habener JF. Stromal cell derived factor-1 (SDF-1)/CXCL12 attenuates diabetes in mice and promotes pancreatic beta-cell survival by activation of the prosurvival kinase Akt. Diabetes. 2007;**56**:2946-2957

[39] Rutti S, Arous C, Schvartz D, Timper K, Sanchez JC, Dermitzakis E, et al. Fractalkine (CX3CL1), a new factor protecting beta-cells against TNFalpha. Molecular Metabolism. 2014;**3**:731-741

[40] Williams O, Fatima S. Granuloma. Treasure Island (FL): StatPearls; 2022

[41] Khader SA, Rangel-Moreno J, Fountain JJ, Martino CA, Reiley WW, Pearl JE, et al. In a murine tuberculosis model, the absence of homeostatic chemokines delays granuloma formation and protective immunity. Journal of Immunology. 2009;**183**:8004-8014

[42] Chensue SW. Chemokines in innate and adaptive granuloma formation. Frontiers in Immunology. 2013;**4**:43

[43] Seiler P, Aichele P, Bandermann S, Hauser AE, Lu B, Gerard NP, et al. Early granuloma formation after aerosol Mycobacterium tuberculosis infection is regulated by neutrophils via CXCR3-signaling chemokines. European Journal of Immunology. 2003;**33**:2676-2686

[44] Torraca V, Tulotta C, Snaar-Jagalska BE, Meijer AH. The chemokine receptor CXCR4 promotes granuloma formation by sustaining a mycobacteria-induced angiogenesis programme. Scientific Reports. 2017;7:45061

[45] Scott HM, Flynn JL. Mycobacterium tuberculosis in chemokine receptor 2-deficient mice: Influence of dose on disease progression. Infection and Immunity. 2002;**70**:5946-5954

[46] Houben RM, Dodd PJ. The global burden of latent tuberculosis infection: A re-estimation using mathematical modelling. PLoS Medicine. 2016;**13**:e1002152

[47] Delemarre EM, van Hoorn L, Bossink AWJ, Drylewicz J, Joosten SA, Ottenhoff THM, et al. Serum biomarker profile including CCL1, CXCL10, VEGF, and adenosine deaminase activity distinguishes active from remotely acquired latent tuberculosis. Frontiers in Immunology. 2021;**12**:725447

[48] Anuradha R, Munisankar S, Bhootra Y, Kumar NP, Dolla C, Babu S. Malnutrition is associated with diminished baseline and mycobacterial antigen - Stimulated chemokine responses in latent tuberculosis infection. The Journal of Infection. 2018;**77**:410-416

[49] Lighter J, Rigaud M, Huie M, Peng CH, Pollack H. Chemokine IP-10: An adjunct marker for latent tuberculosis infection in children. The International Journal of Tuberculosis and Lung Disease. 2009;**13**:731-736

[50] Juffermans NP, Dekkers PE, Peppelenbosch MP, Speelman P, van Deventer SJ, van Der Poll T. Expression of the chemokine receptors CXCR1 and CXCR2 on granulocytes in human endotoxemia and tuberculosis: Involvement of the p38 mitogen-activated protein kinase pathway. The Journal of Infectious Diseases. 2000;**182**:888-894

[51] Lin Y, Zhang M, Barnes PF. Chemokine production by a human alveolar epithelial cell line in response to Mycobacterium tuberculosis. Infection and Immunity. 1998;**66**:1121-1126

[52] Nouailles G, Dorhoi A, Koch M, Zerrahn J, Weiner J 3rd, Fae KC, et al. CXCL5-secreting pulmonary epithelial cells drive destructive neutrophilic inflammation in tuberculosis. The Journal of Clinical Investigation. 2014;**124**:1268-1282

[53] Olmos S, Stukes S, Ernst JD. Ectopic activation of Mycobacterium tuberculosis-specific CD4+ T cells in lungs of CCR7−/− mice. Journal of Immunology. 2010;**184**:895-901

[54] Chakravarty SD, Xu J, Lu B, Gerard C, Flynn J, Chan J. The chemokine receptor CXCR3 attenuates the control of chronic Mycobacterium tuberculosis infection in BALB/c mice. Journal of Immunology. 2007;**178**:1723-1735

[55] Slight SR, Rangel-Moreno J, Gopal R, Lin Y, Fallert Junecko BA, Mehra S, et al. CXCR5(+) T helper cells mediate protective immunity against tuberculosis. The Journal of Clinical Investigation. 2013;**123**:712-726

[56] Gopal R, Rangel-Moreno J, Slight S, Lin Y, Nawar HF, Fallert Junecko BA, et al. Interleukin-17-dependent CXCL13 mediates mucosal vaccine-induced immunity against tuberculosis. Mucosal Immunology. 2013;**6**:972-984

[57] Aravindhan V, Kevinkumar V, Dhamodharan U, Viswanathan V. Serum

levels of chemokines IP-10, IL-8 and SDF-1 serve as good biomarkers for diabetes-tuberculosis nexus. Journal of Diabetes and its Complications. 2018;**32**:857-862

[58] Podell BK, Ackart DF, Obregon-Henao A, Eck SP, Henao-Tamayo M, Richardson M, et al. Increased severity of tuberculosis in Guinea pigs with type 2 diabetes: A model of diabetes-tuberculosis comorbidity. The American Journal of Pathology. 2014;**184**:1104-1118

[59] Lemos MP, Rhee KY, McKinney JD. Expression of the leptin receptor outside of bone marrow-derived cells regulates tuberculosis control and lung macrophage MHC expression. Journal of Immunology. 2011;**187**:3776-3784

[60] Kewcharoenwong C, Saenwongsa W, Willcocks SJ, Bancroft GJ, Fletcher HA, Lertmemongkolchai G. Glibenclamide alters interleukin-8 and interleukin-1beta of primary human monocytes from diabetes patients against Mycobacterium tuberculosis infection. Tuberculosis (Edinburgh, Scotland). 2020;**123**:101939

Chapter 4

Chemokines in Periodontal Diseases

Figen Öngöz Dede and Şeyma Bozkurt Doğan

Abstract

Periodontal disease is a chronic multifactorial inflammatory disease affecting the tooth-supporting apparatus including the gingiva, alveolar bone, and periodontal ligament caused by specific microorganisms. Periodontal diseases are among the most widespread diseases in humans and are a major public health problem due to complications caused by early tooth loss. The immunoinflammatory responses initiated by periodontopathogens to protect the host against periodontal infection cause the release of various proinflammatory and chemotactic cytokines, i.e., chemokines. Chemokines have been implicated in the immunopathogenesis of periodontal disease and are found in gingival tissue, GCF, plasma, and saliva in periodontal disease. This section aims to summarize the data concerning the role of chemokines in periodontal tissue inflammation.

Keywords: periodontal disease, chemokines, periodontal treatment, gingival tissue, gingival crevicular fluid, saliva

1. Introduction

Chemokines are a family of small (8–11 kDa) molecular weight proteins that can bind specific G-protein-coupled cell surface receptors, which are classified as C, CC, CX3C, and CXC subfamilies based on conserved cysteine residues within the N-terminal [1–3]. There were two families of chemokines functionally characterized by inflammatory processes: (1) the CC and (2) the CXC subgroups [4]. Chemokines and chemokine receptors play a central role in the immune response by providing a significant effect on the migration and activation of leukocytes in response to bacterial infection and by acting on the host to control infections [2, 5, 6].

Chemokines have been implicated in the pathogenesis of many inflammatory diseases, including periodontal diseases [2, 7]. It is stated that periodontal diseases are one of the most common infectious diseases among humans [8]. To date, several studies have analyzed various chemokines in periodontal disease and health using saliva, gingival crevicular fluid (GCF), plasma, and gingival tissue samples or in experimental models with periodontal diseases [9–13]. This section aims to summarize the data concerning the role of chemokines in periodontal tissue inflammation.

2. Periodontal diseases

The periodontium refers to the total of the tissues that support the teeth, including the gingiva, periodontal ligament, cementum, and alveolar bone [14]. Periodontal disease is a chronic multifactorial inflammatory disease that develops with the interactions between the dysbiotic dental plaque biofilm and the host immuno-inflammatory response [15]. Although periodontal pathogens play a fundamental role in the initiation and maintenance of periodontal disease, periodontal tissue damage results from prolonged, excessive, and dysregulated immune-inflammatory responses to bacteria and their effects [15]. Most individuals with periodontal disease are not aware of the progress of periodontal tissue destruction because of delays in the detection and treatment of the infection state due to the lack of pain in periodontal diseases [16]. There are two basic forms of periodontal diseases including gingivitis and periodontitis.

Gingivitis is an inflammatory disease that affects the gingival tissues, caused by the imbalance between microorganism products and the host response [17]. The clinical features of gingivitis include the presence of edema, color and contour changes in the gingiva, bleeding on probing or spontaneously, increase in the amount of GCF [17]. In addition, the destruction of periodontal tissues after gingivitis inflammation is reversible [18].

Periodontitis is a destructive form of periodontal disease that destroys the tooth-supporting apparatus, including the gingiva, alveolar bone, and periodontal ligament caused by specific microorganisms [15]. The clinical feature of periodontitis is the existence of clinical attachment loss as a result of inflammatory destruction of the periodontal ligament and alveolar bone [15]. Periodontitis causes irreversible destruction of periodontal tissues [18]. Periodontitis is one of the public health problems due to early tooth loss, negative effects on aesthetic and chewing functions, adverse effects on quality of life, and negative effects on general health [19].

At the International Workshop for classification of periodontal diseases and conditions in 1999, periodontitis is categorized as chronic and aggressive periodontitis [20]. Chronic periodontitis represents the form of destructive periodontal disease that is generally characterized by slow progression and associated with amounts of plaque and calculus [14, 21]. Aggressive periodontitis is a more destructive form of periodontitis (rapid attachment loss and bone destruction) affecting primarily young individuals, possible familial aggregation of disease and not related to amounts of plaque and calculus, including conditions formerly classified as "early-onset periodontitis" and "rapidly progressing periodontitis" [14, 21]. According to the classification, aggressive and chronic periodontitis are subcategorized in local or generalized forms, depending on the percentage of the tooth-affected sites (above or below 30%) and the severity of attachment loss (slight: 1 or 2 mm, moderate: 3 or 4 mm; severe ≥5 mm) [14]. The classification for periodontitis has been updated in 2017 as the forms of the disease previously recognized as "chronic" or "aggressive," are now grouped under a single category, "periodontitis" [19]. A recent classification of periodontitis is based on severity, complexity, risk of progression, and response to treatment [22]. Diagnosis of periodontitis is based on multiple clinical and radiographic parameters. Accordingly, patients are diagnosed with periodontitis when there are interproximal clinical attachment level (CAL) of ≥2 mm or ≥3 mm at ≥2 non-adjacent teeth, inflammation (bleeding on probing and BOP), and radiographic bone loss [22]. Additionally, periodontitis is characterized based on a multidimensional staging (Stage 1,2,3,4) and grading (Grade A,B, and C) system [23]. Periodontitis is affected by several risk factors, including genetic predisposition, smoking habits, and systemic diseases, which include cardiovascular disease, diabetes, and rheumatoid arthritis [24, 25].

3. Pathogenesis of periodontal diseases

The interaction between microbial dental plaque and the host response is responsible for chronic inflammation in the periodontium [26]. The initiation of periodontal disease is due to bacterial infection [2]. Normally, periodontal tissue is highly responsive to oral microbial stimulation with the coordinated release of host defense mediators [24]. Excessive pathogenic bacterial invasion into periodontal tissues disrupts the immune response and causes the release of excessive inflammatory mediators in tissues that will destroy periodontal tissues [24, 25]. İnflame periodontal tissue includes an accumulation of T cells, B cells, macrophages, and dendritic cells [3].

It has been proven that not all bacteria adhered to the tooth surface, but some pathogenic bacteria in the biofilm cause periodontal disease [2]. The three main commonalities of *Actinobacillus actinomycetemcomitans (A.a.), Tannerella forsythia (T. forsythia), and Porphyromonas gingivalis (P. gingivalis)* have features, including gram-negative, produce lipopolysaccharide, modulate the local inflammatory response in host cells, and invasion ability to inside of the mucosal barrier and epithelial cells [21]. *Porphyromonas gingivalis (P. gingivalis)* is the dominant oral pathogen associated with periodontitis [16]. *P. gingivalis* expresses three major virulence factors: fimbriae, gingipains, and lipopolysaccharides [27]. *P. gingivalis* also inhibits neutrophil chemotaxis and may inhibit the influx and activation of monocytes/macrophages leading to an overall reduction in innate immunity [5].

The inflammatory and immune responses, initiated by periodontopathogens, are thought to protect the host against infection [28]. However, the host immunoinflammatory response to the bacterial biofilm in periodontitis leads to the release of several proinflammatory and chemotactic cytokines, that is, chemokines [25]. Chemokines can be secreted from cells of the periodontium, such as fibroblasts, endothelial cells, and epithelial cells, in response to bacterial load [29]. Chemokines have been implicated in the immunopathogenesis of periodontal disease, and are found in gingival tissue, GCF, and saliva in periodontal disease [28, 30, 31].

4. Chemokines in periodontal disease

4.1 IP-10\CXCL10 (receptor CXCR3 ligand) and Periodontal Disease

C-X-C motif chemokine ligand 10 (CXCL10), known also as interferon gamma-induced protein 10 (IP-10), is a member of the CXC chemokine family, and acts as a chemoattractant for several cells, such as monocytes/macrophages, T cells, NK cells, and dendritic cells, but not neutrophils [1, 26]. CXCL10 also plays an important role in leukocyte homing to inflamed tissues [32]. IP-10 is a ligand for CXCR3 receptors on Th1 cells [33]. It has been shown in many studies to be involved in tissue destruction in periodontal disease [26, 33, 34].

Gemmell et al. [5] reported that keratinocyte expression of IP-10 decreased with increased inflammation. Some reports determined that IP-10 and its receptor CXCR3 expressions in gingival tissues were more abundant and higher in patients with aggressive periodontitis and marginal periodontitis [33, 35]. It has been detected that the average ratio of CXCR3-expressing T cells in inflamed gingival tissues of patients with marginal periodontitis is in the range of about 0.8 and 4.5% [33]. Moreover, a recent animal study found that CXCL10 expressions were significantly upregulated in the gingival biopsies of the rats with experimental periodontitis compared to healthy controls, and also in periodontal fibroblasts exposed to the periodontopathogen *F. nucleatum* [1]. Another study demonstrates PTM of CXCL10 by gingipains of *P. gingivalis* and that strain differences may particularly affect the activity of these bacterial membrane-associated proteases [36].

Sakai et al. [37] determined that GCF IP-10 levels significantly increased in patients with chronic periodontitis compared to healthy periodontal individuals. Furthermore, Shimada et al. [34] reported that IP-10 levels in GCF were significantly higher in disease sites than in healthy sites, and in BOP-positive diseased sites compared to BOP-negative diseased sites of the patients with generalized chronic periodontitis. It was also determined that there were significant correlations between GCF IP-10 levels and the *P. gingivalis* ratio [34]. Peyyala et al. [38] found that *P. gingivalis* and oral streptococcal biofilms inhibited IP-10 production and IP-10 levels increased more against biofilm in their study using a bacterial biofilm model to stimulate oral epithelial cells. On the other hand, Thunell et al. [26] obtained GCF samples from the diseased and healthy sites of patients with generalized severe chronic periodontitis 6–8 weeks after non-surgical periodontal treatment and found that the IP-10 levels significantly increased in the healthy sites after treatment, but there was a nonsignificant increase in the diseased sites. The elevation of IP-10 after therapy was associated with an overall decrease in inflammatory and disease parameters; therefore, it may have a role in wound healing processes rather than reflecting tissue destruction [26]. Conversely, another study determined that calcitriol treatment (the biologically active form of vitamin D) in periodontal lesions significantly reduced CXCL10 production in IL-1β-stimulated human periodontal ligament cells (HPDLC) [39].

Aldahlawi et al. [9] demonstrated that CXCL10 levels in saliva and serum significantly increased in the patients with chronic periodontitis compared with periodontally healthy controls, and there was a significant positive correlation between the clinical parameters of periodontal disease and CXCL10. In addition, serum CXCL10 level was significantly higher in the moderate to severe periodontitis group, which is defined by deeper PD and worse CAL, than the mild periodontitis group [9]. Furthermore, the serum CXCL10 was higher in older subjects (>30 years old) who had significantly more attachment loss, than younger subjects (<30 years old) [9]. Likewise, Panezai et al. [40] reported that serum CXCL10 was positively associated with the number of teeth and some inversely related to MBL (marginal bone loss). It has been indicated that CXCL10 might modulate the pathogenesis of periodontal disease, thus making it useful as a diagnostic biomarker [9, 33].

4.2 MCP-1\CCL2 (receptor CCR4) and periodontal disease

The chemokine (C-C motif) ligand 2 (CCL2) also referred to as monocyte chemoattractant protein 1 (MCP-1), is a strong chemoattractant for monocytes, lymphocytes, natural killers, and macrophages [5, 24]. MCP-1 attracts CCR2- and CCR4-positive cells and is linked to Th2 responses [13, 35]. Several signaling pathways involved in the increased MCP-1 in inflammatory responses include the NF-κB pathway, TLR2/4 signaling pathway, phosphatidylinositol 3-kinase/Akt pathway, and MAPK signaling pathways [24]. It has been argued that in particular, the MAPK signaling pathway can determine the role of MCP-1 in periodontal diseases, as the MAPK signaling pathway increases MCP-1 production in human gingival fibroblasts [41]. It has been stated that Gram-negative bacterial LPS, which is one of the most important causes of periodontal diseases, can also activate the MAPK pathway in periodontal tissue cells [24]. However, it showed that MCP-1 could not provide an adequate signal in the epithelial cell response against oral biofilm [38].

In the past, an immuno-histochemistry study did not only demonstrate a high level of MCP-1 in human inflamed gingival tissues but also a significantly higher MCP-1 gene expression in patients with chronic periodontitis [42]. In a study, while MCP-1 gene expression was determined in the gingival tissue of adult periodontal patients, it could not be detected in healthy controls, and they strongly suggested that MCP-1 may play an important role in monocyte infiltration in periodontal tissues of periodontal patients [43]. Another study stated that MCP-1 is present in human inflamed gingival tissue of marginal periodontitis and is responsible for modulating the disease process [33]. Similarly, it was determined that MCP-1 and its receptor CCR4 expressions in gingival tissues were more abundant and higher in patients with chronic periodontitis [35]. A recent study investigating inflamed and healthy periodontal tissue from intrabony periodontal lesions determined that MCP-1/CCL2 expression levels were higher in inflamed tissue compared to healthy periodontal tissue [32]. Besides, an *in vitro* study found that CCL2 increased the number of M2 (alternatively activated, anti-inflammatory) phenotype macrophages and decreased TNF-α secretion. And manipulation of endogenous M2 (alternatively activated) phenotype macrophages with CCL2 controlled-release microparticles (MPs) decreased the M1(classically activated, pro-inflammatory) phenotype, and M2 phenotype ratio and prevented alveolar bone loss in mouse periodontitis models [25]. Authors stated that the delivery of CCL2 MPs provides a novel approach to treatment of periodontal diseases [25].

Gemmell et al. [5] reported that keratinocyte expression of MCP-1 decreased with the increased inflammation. Furthermore, Tonetti et al. [44] investigated in situ expression of MCP-1 mRNAs in human periodontal infections, and determined that MCP-1 was expressed in the chronic inflammatory infiltrate and along the basal layer of the oral epithelium. An animal study found that MCP-1 in periodontal ligament cells (PDL) showed significantly higher expression in the periodontitis rats group than in the control group, and also mRNA levels of chemokines MCP-1 were significantly upregulated [24]. Furthermore, Nebel et al. [45] determined that the expression of CCL2 in human PDL cells was higher at both mRNA and protein levels than that of the CCL3 chemokine, and stated that PDL cells can produce high amounts of CCL2. Souto et al. [6] found it to be positively correlated with increased densities of CD1a+ dendritic cells (DCs) and CCL2 expression in gingival tissue of patients with chronic periodontitis. Furthermore, an *in vitro* study showed that MCP-1 under LPS stimulation of *P. gingivalis* and *A. Actinomycetemcomitans* was found secreted at higher levels in bone-derived cells (especially osteoblastic cells) and at more moderate levels in mononuclear cells [46]. Another *in vitro* study investigating the potential effect of sCD14 on the HPDLSC response to two different TLR-2 agonists reported finding CCL2 production at higher sCD14 levels.

Hanioka et al. [47] determined that the substance P (SP) level in GCF showed a significant correlation with MCP-1 in patients with slightly or moderately advanced periodontitis. Besides, Bamashmous et al. [48] stated that MCP-1/CCL2 levels in the GCF of individuals with experimental gingivitis were very low and yet contribute to the normal bone turnover process or inflammatory bone loss in periodontitis. Gupta et al. [49] determined that MCP-1 levels in saliva, serum, and GCF of individuals with chronic periodontitis were significantly higher than in the control group, and these levels decreased significantly after non-surgical periodontal treatment in individuals with chronic periodontitis, and there were significant positive correlations among the levels of MCP-1 in GCF, saliva, serum, and clinical parameters. Additionally, Pradeep et al. [11, 12, 50] found that MCP-1 levels in serum and GCF were higher in the chronic periodontitis group than in the gingivitis and control groups, and also in the gingivitis group than in the control group. They also determined that GCF and serum MCP-1 levels decreased after non-surgical periodontal treatment in the periodontitis group, and positively correlated with clinical parameters [11, 12, 50]. Another study analyzed that MCP-1 was detected in chronic periodontitis and gingivitis sites, especially in severely inflamed sites, but was not detectable in periodontally healthy sites, and found that MCP-1 concentrations in GCF were significantly higher in chronic periodontitis sites than in gingivitis sites [51]. Authors also stated that there was a significantly positive correlation between *B. forsythus* amount and MCP-1 [51]. Likewise, Thunell et al. [26] examined GCF samples from the diseased and healthy sites of patients with generalized severe chronic periodontitis 6–8 weeks after non-surgical periodontal treatment and determined that the MCP-1 levels significantly decreased in the diseased sites after treatment. Unlike, Silva et al. [52] found that no significant difference was detected in the total amount and concentration of MCP-1 in GCF in the active and inactive regions of patients with moderate and severe chronic periodontitis. Similarly, no significant relationship was found in CCL2\MCP-1 levels in the gingival tissue and GCF of smokers and nonsmokers with chronic periodontitis and healthy group [4, 53]. The authors only found that serum MCP-1 levels were higher in smokers with periodontitis than nonsmokers [4]. Conversely, in another study examining the effect of smoking on MCP-1 levels, it was detected that GCF MCP-1 levels in the diseased sites of nonsmokers with chronic periodontitis were significantly higher when compared to smokers with disease and control groups [54]. Moreover, the authors stated that MCP-1 levels decrease in smokers, and this decrease may be caused by disruptions in neutrophilic chemotaxis and migration [54]. Hereby, a recent meta-analysis emphasized that GCF levels of MCP-1/CCL2 were significantly higher in patients with chronic periodontitis than in periodontal healthy controls, and decreased after non-surgical periodontal treatment [7].

Unlike Gupta et al. [49], Kawamoto et al. [13] found that MCP-1/CCL2 levels in saliva were significantly reduced in the patients with an incisor-molar pattern of the rapid rate of progression compared to healthy controls, but no significant difference was found between the Stage III periodontitis patients and healthy controls.

Martins et al. [55] determined that although there was no significant difference in the GCF MCP-1 levels of individuals with localized (LAgP) and generalized aggressive periodontitis (GAgP) when compared with the control group at baseline, their levels increased both compared to the baseline levels intra-group and compared to the control group after non-surgical periodontal treatment. When the study also examined serum MCP-1 levels, they determined that pre-and post-treatment levels of LAgP and GAgP patients were significantly higher in the control group [55]. Similarly, Shaddox et al. [56] recognized that MCP-1 levels in GCF were increased in healthy sites compared with diseased sites in the patients with LAgP. By contrast, Emingil et al. [10] showed that GCF MCP-1 levels were elevated in the patients with GAgP compared to the healthy group and that there was a significant positive correlation between GCF MCP-1 and both probing depth and clinical attachment loss.

Kurtis et al. [57] found that MCP-1 levels in GCF were higher in both patients with chronic and aggressive periodontitis compared to healthy controls, but no statistical difference was found between the two types of periodontitis. In line with the results of the study by Emingil et al. [10], they determined that MCP-1 in GCF had positive correlations with periodontal clinical parameters [57].

Previous studies showed that the expression levels of MCP-1/CCL2 were increased with the progress of periodontitis and thus indicated to be the major chemoattractant of macrophages in periodontal diseases [4, 24].

4.3 MCP-3\CCL7 and periodontal disease

The chemokine (C-C motif) ligand 7 (CCL7) also referred to as monocyte chemoattractant protein-3 (MCP-3), is a powerful chemotactic protein expressed by endothelial cells and monocytes and included Th2 cell chemoattractants [2, 52].

Dezerega et al. [2] determined that MCP-3 levels in GCF were higher in the chronic periodontitis group compared to the control group, and the total amount of MCP-3 per site was significantly higher in active sites than inactive sites of the patients with chronic periodontitis. Authors also found that MCP-3 expression in gingival tissue of the patients with chronic periodontitis was localized to inflammatory cells, especially plasmocytes and vascular endothelium, but MCP-3 was not detected in healthy controls [2]. It has been argued that raised levels of MCP-3 were involved in inflammatory cells in periodontal tissues and may be associated with the initiation and progression of periodontal diseases [2].

4.4 MIP-1alpha\CCL3 (its receptor CCR5) and periodontal disease

Chemokine (C-C motif) ligand 3 (CCL3) also known as macrophage inflammatory protein 1-alpha (MIP-1-α) is a potent chemoattractant for monocytes, lymphocytes, and macrophages [5]. High levels of MIP-1α are produced by osteoblasts and MIP-1α expression has been linked to bone remodeling [46] and acts to stimulate osteoclasts [6]. The CCL3 chemokine is a protein associated with important biological phases of bone remodeling [53]. MIP-1α/CCL3, a chemokine associated with bone homeostasis, is completely shut down during experimental gingivitis, indicative of a significant alteration in bone turnover processes and an important biomarker of periodontitis [48]. In the regression models, MIP-1α was the biomarker that best discriminated periodontal disease from health compared with OPG, ICTP, and b-CTX [58].

CCL3 has a potential role in inflammatory bone resorption in the periodontal environment [59]. Indeed, CCL3 positive cells increase in number with increasing severity of periodontal disease and are associated with an augmented proportion of lymphocytes in inflamed tissues [30]. Moreover, a study decided that keratinocyte expression of MIP-1 was more abundant in diseased periodontal tissue, and suggested that it plays a role in recruiting leukocytes through the epithelium in both the early and late stages of inflammation [5]. An *in vitro* study found that MIP-1α under LPS stimulation of *P. gingivalis* and *A. Actinomycetemcomitans* secreted at higher levels in mononuclear cells, but did not enhance MIP-1α production in osteoblastic cell populations [46].

Garlet et al. [35] determined that expressions of MIP-1α and its respective receptor CCR5 in gingival tissue were more intense in both periodontitis groups compared to the control group, and also higher in the aggressive periodontitis group compared to the chronic periodontitis group. A study examining the levels of MIP-1/CCL3 in inflamed and healthy periodontal tissues of patients with periodontitis found that it was expressed only in inflamed tissues and that MIP-1/CCL3 was associated with the acute phase of inflammation as macrophage-secreting cytokines [32]. It has been found that MIP-1/CCL3-producing cells were detected in all of the samples of inflamed gingival tissues. On the other hand, CCR5-positive cells were detected in all of the inflamed and healthy periodontal tissue samples, although they were found in large numbers in the inflamed gingival tissues [33]. Moreover, Souto et al. [6] revealed that CCL3 levels in gingival tissue were raised in the patients with mild-moderate and advanced chronic periodontitis compared with healthy subjects, but no difference in both periodontitis groups, and the percentage of CAL>3mm sites and CCL3 levels were positively correlated.

Bamashmous et al. [48] revealed that MIP-1α/CCL3 levels in the GCF of individuals with experimental gingivitis were significantly reduced in all three clinical response groups at the first gingivitis measurement (day 4) and were restored at the first time point in the resolution phase (day 28). Haytural et al. [4] found that MIP-1α levels in GCF increased in smokers and nonsmokers with chronic periodontitis compared to healthy groups, but no difference was observed between the smokers and the nonsmokers with chronic periodontitis. Conversely, the authors revealed that serum MIP-1α levels were higher in healthy nonsmokers than in nonsmokers with chronic periodontitis, and there was no significant difference in smokers [4]. Thus, it has been suggested that MIP-1α plays an important role in periodontal inflammation [4]. On the other hand, studies examining the effect of smoking on GCF and gingival tissue CCL3 levels have shown that smokers with chronic periodontitis had lower levels compared to nonsmokers with chronic periodontitis [53, 54]. Thunell et al. [26] examined GCF samples from the diseased and healthy sites of patients with generalized severe chronic periodontitis 6–8 weeks after non-surgical periodontal treatment and determined that MIP-1α levels significantly decreased in diseased sites after treatment. It was found that GCF MIP-1α levels were significantly higher at 3 and 6 months after non-surgical periodontal treatment compared to baseline. Emingil et al. [60] stated that the chemokine activity would account for the regulation of the inflammatory response to subantimicrobial-dose doxycycline therapy (SDD; the only Food and Drug Administration–approved host-modulation therapy).

Salivary levels of MIP-1α proved to be significantly increased in patients with chronic periodontitis (18-fold) compared to healthy controls, and demonstrated a strong correlation with the clinical parameters of periodontal diseases, such as BOP, probing depth (PD) ≥ 4 mm, PD ≥ 5 mm, and percentage of CAL [58]. Fine et al. [59] found that MIP-1α levels in saliva were significantly elevated in the *A. Actinomycetemcomitans* + LAgP (50-fold) group 6 to 9 months before the detection of bone loss compared with the *A. Actinomycetemcomitans* + LAgP group and increasing levels of MIP-1α correlated with increasing PD. However, Kawamoto et al. [13] stated that saliva CCL3/MIP-1α levels in the periodontitis group (Stage III and IV) when compared with the control groups could not find a significant difference. Nevertheless, studies have displayed that MIP-1α levels in saliva can be used as a biomarker for the detection of progressive bone loss [58, 59].

4.5 RANTES\CCL5 and periodontal disease

Chemokine (C-C motif) ligand 5 (CCL5) also known as RANTES (regulated on activation, normal T-cell expressed and secreted) is a potent chemoattractant for the Th1 cells with no effect on the Th2 cells [4].

Previous studies found that the levels of RANTES in GCF of individuals with periodontitis raised compared to the control group, and in addition, it was significantly higher in active sites than inactive sites in the periodontitis group [34, 61–63]. Moreover, periodontal treatment reduced GCF RANTES levels in patients with chronic periodontitis [26, 61, 63]. Haytural et al. [4] found that RANTES levels in GCF were increased in smokers and nonsmokers with chronic periodontitis compared to healthy groups, but no difference between smokers and nonsmokers with chronic periodontitis, and also no significant difference was in the levels of serum RANTES between smokers and nonsmokers with chronic periodontitis and healthy groups. This effect may be due to impaired vascularization as a function of smoking and disrupted inflammatory processes [4]. Besides, Tymkiw et al. [54] determined that GCF RANTES levels in the diseased sites of nonsmokers with chronic periodontitis were significantly higher when compared to smokers with disease and control groups and indicated that smoking suppressed GCF RANTES levels.

Emingil et al. [10] showed that GCF RANTES levels raised in the patients with generalized aggressive periodontitis compared to the healthy group and that there was a significant positive correlation between GCF MCP-1 and both probing depth and clinical attachment loss, but no correlation between GCF RANTES levels and the percentage of sites with bleeding. Another study determined that GCF RANTES levels elevated after non-surgical periodontal treatment with and without adjunctive SDD groups compared to baseline, however, there was no difference between groups with and without SDD adjunctive [60].

Gemmell et al. [5] reported that keratinocyte expression of RANTES decreased with increased inflammation. Lee et al. [32] investigated inflamed and healthy periodontal tissues obtained from intrabony periodontal lesions and determined that RANTES/ CCL5 expression levels were higher in inflamed tissue than in healthy periodontal tissues and RANTES/ CCL5 appeared to play a role in the migration of hPDLSCs (human periodontal-ligament stem cells) into inflammatory periodontal lesions. Another study stated that RANTES-producing cells were not been found in gingival tissues in patients with marginal periodontitis [33]. Increased RANTES/ CCL5 levels were shown in whole blood cell cultures (WBCC) stimulated with LPS of the patients with periodontitis compared with the control group and also, and these levels did not change after non-surgical periodontal therapy [64]. Repeke et al. [30], in their study, carried out the experimental periodontal disease with Aggregatibacter actinomycetemcomitans-infected C57Bl/6 (WT) in mice, found that a significant reduction of experimental periodontitis is verified through the treatment with met-RANTES (a CCR1 and CCR5 antagonist). Furthermore, an *in vitro* study determined that RANTES production by LPS stimulation of *P. gingivalis* and *A. actinomycetemcomitans* induced moderate levels in both mononuclear and osteoblastic cells [46].

4.6 IL-8\CXCL8 and periodontal disease

Interleukin 8 (IL-8 or chemokine (C-X-C motif) ligand 8, CXCL8), the first cytokine identified to have chemotactic activity, is a potent neutrophil chemoattractant and activator of human neutrophils via interaction with two receptors (CXCR1 and CXCR2) [65]. IL-8 is involved in the initiation and amplification of acute inflammatory reactions; it is secreted by several cell types in response to inflammatory stimuli [65]. IL-8/CXCL8 has a direct effect on osteoclast differentiation and activity by signaling through the specific receptor, CXCR1 [52]. A meta-analysis reported that there was evidence of higher levels of IL-8 in individuals with chronic periodontitis compared with periodontally healthy controls [7].

A previous study suggested that patients with chronic periodontitis had a subpopulation of peripheral neutrophils with higher responsiveness to IL-8 priming than the control group [31]. An *in vitro* study investigated the potential effect of sCD14 on the hPDLSC response to two different TLR-2 agonists and determined that the production of CXCL8 was gradually increased by both TLR-2 agonists and was significantly enhanced by sCD14 [27]. Tonetti et al. [44] analyzed in situ expression of IL-8 mRNAs in human periodontal infections, and IL-8 expression in gingival tissue was maximal in the junctional epithelium adjacent to the infecting microorganisms. Moreover, a recent animal study showed that mRNA levels of IL-8 were significantly upregulated in LPS-stimulated periodontal ligament (PDL) cells in rats with experimental periodontitis [24]. It was observed that IL-8 was expressed only in inflamed tissues from the inflamed and healthy periodontal tissues obtained from intrabony periodontal lesions [32]. Another animal study confirmed to secrete IL-8 in PDL cells after LPS stimulation, and significantly upregulated in the periodontitis rats group compared with the control group [24].

Souto et al. [6] found that CXCL8 levels increased in the gingival samples of patients with chronic periodontitis compared with healthy mucosa and a positive correlation was observed between CXCL8 in the gingival tissue and CAL >3 mm, and other studies obtained that smoking reduced CXCL8 levels in gingival tissue of chronic periodontitis patients [53].

GCF IL-8 levels decreased or did not change in the experimental gingivitis after 4 weeks of plaque accumulation compared to the control group [48, 66]. On the other hand, previous studies [62, 63] determined that IL-8 levels in GCF were higher in patients with moderate to advanced periodontitis than in the control group, and in active sites than in inactive sites of periodontitis patients. Also, periodontal therapy reduced GCF IL-8 levels in periodontitis patients. Similarly, Thunell et al. [26] found that reassessment 6–8 weeks after initial periodontal treatment reduced GCF CXCL8/IL-8 levels in the diseased sites of the patients with generalized severe chronic periodontitis.

There was no statistically significant difference between salivary IL-8 levels of individuals with Stage III periodontitis both moderate (GB) and incisor-molar patterns of the rapid rate of progression (GC/IMP) [13]. Conversely, plasma IL-8 levels increased in both type 1 diabetes mellitus (T1DM) and type 2 diabetes mellitus (T2DM) patients with chronic periodontitis compared to the systemic healthy chronic periodontitis and control group. It was argued that increased plasma IL-8 levels may be associated with the presence of both types of diabetes mellitus in periodontal disease [65]. In addition, no significant relationship between the periodontopathic bacteria and IL-8 production was found [65]. Furthermore, Lappin et al. [67] detected more increased plasma IL-8 levels in the T1DM+chronic periodontitis group than alone in chronic periodontitis and control groups and alone in chronic periodontitis group in than the control group. Also, they found a correlation between IL-8 plasma levels and HbA1c, PD, and attachment loss. Another similar study determined increased IL-8 levels in blood samples stimulated with and without *P. gingivalis* and *Escherichia coli* LPS of chronic periodontitis patients with and without T2DM [68]. Similarly, Mohamed et al. [69] reported significantly elevated levels of IL-8 in GCF in patients with T2DM+chronic periodontitis as compared to the alone chronic periodontitis group. Conversely, Engebretson et al. [70] stated statistically decreased levels of IL-8 in GCF of chronic periodontitis patients with T2DM compared to those without diabetes mellitus with chronic periodontitis. It was stated that there was no significant difference in GCF IL-8 levels between diabetes with and without chronic periodontitis and nondiabetes periodontitis groups [71]. The bidirectional relationship between periodontal disease and diabetes has been scientifically proven and IL-8 is an important predictor of this relationship [67, 69].

In the past, IL-8 production remained unchanged before and after periodontal treatment *in Escherichia coli* LPS-stimulated whole blood cell cultures of patients with periodontitis [63]. Afterward, studies evaluating the effect of vitamin D reported that *P. gingivalis* significantly promoted the protein expressions of IL-8 in hPDLCs, and both 1,25 dihydroxy vitamin D3 and 25-hydroxyvitamin D3 combined with *P. gingivalis* inhibited the protein expression of IL-8 compared with *P. gingivalis* treatment alone [72, 73]. Thus, the authors indicated that vitamin D may potentially inhibit the periodontal inflammation induced by *P. gingivalis* partly by decreasing the IL-8 expression in hPDLCs [72, 73]. Moreover, Hosokawa et al. [39] stated that the production of IL-8 in IL-1β-stimulated HPDLC significantly raised after calcitriol treatment in chronic periodontitis patients.

4.7 SDF-1\CXCL12 (receptor CXCR4) and periodontal disease

Stromal-derived factor-1 (SDF-1α and β), also known as CXC chemokine ligand 12 (CXCL12), is a potent chemoattractant, which was originally isolated from a murine bone marrow stromal cell line [8]. The interaction of SDF-1/CXCL12 with the receptor, CXCR4, which is expressed in human osteoclast precursors, induces chemotaxis and differentiation into osteoclasts [28]. In intraosseous periodontal lesions, the expression of CXCL12/SDF-1 was lower or absent in inflamed tissues compared to healthy tissues. Therefore, it is named as decreasing/disappearing chemokine group, known to be homeostatic chemokines, although they could also be involved in inflammatory reactions [32]. In addition, Hosokawa et al. [74] exhibited that CXCL12 and CXCR4 mRNA were expressed in both normal gingival tissues and periodontal disease tissues. Additionally, the authors found that TNF-α, IFN-γ, and TGF-β1 increased CXCL12 production by HGF and decreased CXCR4 expression by HGF [74]. The study also stated that *P. gingivalis* LPS diminished the CXCL12 production and CXCR4 expression by HGF, and indicated that it may be associated with the progression of periodontal disease [74].

A previous study found that SDF-1α levels in GCF and gingival tissue were higher in the chronic periodontitis group than healthy group and these levels were reduced after non-surgical periodontal therapy in the periodontitis group [8]. In addition, the authors remarked that the presence of SDF-1 increases neutrophil migration and is involved in immune defense in periodontal disease, and thus may play a role in the development of periodontal disease and be a useful biomarker in its determination [8]. Otherwise, no statistical difference was the levels of salivary CXCL12/SDF-1α between moderate and severe Stage III periodontitis groups and their controls [13].

4.8 MCP-2\CCL8 and periodontal disease

Chemokine (C-C motif) ligand 8 (CCL8), also known as monocyte chemoattractant protein-2 (MCP-2), was observed during inflammatory response for its monocyte and T-lymphocyte attractant properties [75]. It was stated that the levels of CCL8 increased significantly in the periodontal ligament 24 hours after orthodontic tooth movement, both *in vivo* and *in vitro* [75]. Authors also reported that CCL8 decreased OPG mRNA but did not significantly increase RANKL expression, and it was suggested that CCL8 could induce positive effects on osteoclastogenesis [75]. Similarly, Oliveira et al. [76] detected that CCL8 levels in gingival tissue increased in both with and without diabetic rats with periodontal disease compared to its respective control group, and decreased after Alisk treatment. However, a recent study found that there was no statistical difference in the levels of salivary CCL8/MCP-2 between patients with periodontitis in Stage III classified as moderate (GB) or incisor-molar pattern of a rapid rate of progression (GC/IMP) and their respective healthy controls [13].

4.9 CCL20\MIP-3α (receptor CCR6) and periodontal disease

Chemokine (C-C motif) ligand 20 (CCL20), also known as macrophage inflammatory protein-3 (MIP-3α) is produced by the epithelial cells of inflamed epithelial tissues and is the most potent chemokine for the selective attraction of immature DCs *in vitro* through an interaction with the CCR6 receptor [48, 53]. MIP-1α/CCL3, a chemokine associated with bone homeostasis is completely shut down during experimental gingivitis, indicative of a significant alteration in bone turnover processes [48]. An experimental gingivitis study found that MIP- 3α/CCL20 levels were significantly higher in the high response group than in the low response group [48]. Furthermore, Souto et al. [6] detected that CCL20 levels in gingival tissue were increased in the individuals with advanced chronic periodontitis compared with mild-moderate chronic periodontitis, but no statistical difference was observed between the chronic periodontitis and the control groups. Authors also indicated that CCL20 plays a more important role in the migration of DCs in advanced stages of chronic periodontitis [6]. A study examining the effect of smoking found that smoking did not affect CCL20 levels in the gingival tissue of individuals with chronic periodontitis [53]. Another study found that there was no statistical difference in the levels of salivary CCL20/MIP-3α between patients with periodontitis in Stage III classified as moderate (GB) or incisor-molar pattern of a rapid rate of progression (GC/IMP) and their respective healthy controls [13]. On the other hand, a therapeutic study stated that treatment of calcitriol significantly inhibited CCL20 production in IL-1β-stimulated HPDLC and argued that vitamin D could inhibit bone destruction in periodontal lesions by suppressing CCL20 production [39].

4.10 CXCL5\ENA-78 and periodontal disease

C-X-C motif chemokine 5 (CXCL5), also known as epithelial-derived neutrophil-activating peptide-78 (ENA-78), binds to the CXCR2 and stimulates the chemotaxis and activation of neutrophils [1]. Moreover, it is involved in angiogenesis and connective tissue remodeling well as cancer cell proliferation, migration, and invasion [1].

A recent study noted that CXCL5 levels in the periodontal cells of the extracted tooth were significantly upregulated in the inflamed group compared to the healthy controls, and the number of CXCL5 positive cells was higher in the inflamed group, especially in the epithelial layer [1]. At the same time, the authors supported these data by finding that CXCL5 was significantly upregulated in the gingival tissue of rats with experimental periodontitis [1]. It was emphasized that CXCL5 was an important molecule in the pathogenesis of periodontal diseases [1]. Furthermore, increased levels of CXCL5 have been detected in gingival tissues of experimentally-induced periodontitis in rodents [77]. Another animal study determined that CXCL5 expression in gingival tissue increased in the wild-type *P. gingivalis*-infected group compared to knockout group^{-}\MMP-8^{-} and, suggested that MMP-8 is associated with a reduced expression of CXCL5 in the *P. gingivalis*-induced experimental periodontitis model [78]. On the other hand, there was no significant difference in salivary CXCL5/ENA-78 levels when compared with individuals with moderate to severe periodontitis and their respective control groups [13].

4.11 CXCL16\SCYB16 (its receptor CXCR6) and periodontal disease

CXC ligand (CXCL) 16 is a chemokine identified in DCs, endothelial cells, B cells, T cells, smooth muscle cells, and macrophages [3, 79].

Hosokawa et al. [3] detected CXCL16 and CXCR6 mRNA expression in both healthy and diseased periodontal tissue, but it was significantly more intense in diseased tissue compared to healthy tissues. In diseased tissue, CXCL16 was strongly expressed by unstimulated human gingival fibroblasts (HGFs), and CXCR6-positive cells that were generally distributed near the sulcular epithelium, where the initial bacterial challenge to the host occurs in periodontitis [3]. Moreover, while CXCL16 mRNA expression upregulated stimulation with pro-inflammatory cytokines IL-1β, TNF-α, and IFN-γ, the expression of CXCL16 by HGFs in periodontal tissues was inhibited by IL-4 and IL-13 produced in Th2 cells [3]. It was suggested that the CXCL16 produced by HGFs in diseased periodontal tissue may play a role in the attraction of T cells to diseased tissue and the exacerbation of periodontal disease [3]. An experimental gingivitis study found that SCYB16/CXCL16 levels were significantly higher in the high response group compared to the low response group [48]. On the other hand, CXCL16 levels were examined in the saliva of patients with Stage III periodontitis (both moderate and severe) and no significant difference was found when compared with control subjects [13].

A correlation study determined that there was no significant relationship between clinical periodontal parameters and plasma CXCL16 levels [79]. Multiple regression analyses revealed that CXCL16 levels in plasma were significantly related to smoking and associated with more severe periodontitis, especially PD ≥7 mm and clinical AL ≥5 mm [79].

4.12 CCL19\MIP-3β and periodontal disease

Chemokine (C-C motif) ligand 19 (CCL19), also known as EBI1 ligand chemokine (ELC) and macrophage inflammatory protein-3-beta (MIP-3β), is expressed abundantly in thymus and lymph nodes and binds to the CCR7 [53]. Souto et al. [6] detected increased CCL19 levels in gingival tissue in the individuals with advanced chronic periodontitis compared with healthy tissue, but no differences could be observed when comparing mild-moderate and advanced chronic periodontitis groups. In a study, even though CCL19 in gingival tissue did not reveal a statistically significant decrease in smokers with chronic periodontitis compared to nonsmokers, negative correlations could be observed between CCL19 levels and time of the smoking habit in years (SH/years) [53]. Authors suggested that the correlation between CCL19 and SH/years supports the notion that the negative effects of smoking on periodontal health also appear to be dose-related [53].

4.13 CCL25\TECK and periodontal disease

Chemokine (C-C motif) ligand 25 (CCL25), also known as TECK (thymus-expressed chemokine), is chemotactic for thymocytes, macrophages, and DCs and binds to the CCR9 [13]. Kawamoto et al. [13] implicated that CCL25 levels in saliva decreased in patients with Stage III periodontitis as an incisor-molar pattern of a rapid rate of progression compared to controls, but no differences could be observed when analyzing moderate Stage III periodontitis and control groups.

4.14 CCL17 or TARC and periodontal disease

Chemokine (C-C motif) ligand 17 (CCL17), also known as TARC (Thymus and activation regulated chemokine), is a powerful chemokine produced in the thymus and by antigen-presenting cells like DCs, macrophages, and monocytes and binds to the CCR4 [28]. It is a Th2 cell chemoattractant [28]. It was suggested that the expression of Th2 and Treg chemoattractants (TARC/CCL17) could attenuate periodontal disease severity [28]. Kawamoto et al. [13] implicated that CCL17 levels in saliva increased in patients with incisor-molar pattern Stage III periodontitis compared to controls, but no differences could be observed when analyzing moderate Stage III periodontitis and control groups.

4.15 CCL27\CTACK and periodontal disease

C-C motif chemokine ligand 27 (CCL27), also known as CTACK (cutaneous T-cell-attracting chemokine) is expressed in numerous tissues, including gonads, thymus, placenta, and skin and binds to the CCR10 [13]. A recent study implicated that CCL27 levels in saliva increased in patients with incisor-molar pattern Stage III periodontitis compared to controls, but no differences could be observed when analyzing moderate Stage III periodontitis and control groups [13].

5. Conclusions

Periodontal diseases are a common public health problem due to the lack of pain, the fact that patients do not realize their periodontal tissue loss, late access to treatment, sometimes severe and aggressive periodontal tissue destruction at younger ages, and the difficulty of treatment. Progression of periodontitis, an irreversible form of periodontal disease, causes early tooth loss, and this leads to problems that impair chewing function, aesthetics, social inequality, and quality of life in the patient. Clinically undetectable periodontal diseases are determinable in the early period with the levels of biomarkers to be examined in biological fluids. Many studies have analyzed different chemokines and chemokine receptors in biological fluids in human studies and experimental models in periodontal disease and health. In fact, the role of chemokines in periodontal disease was explained by *in vitro* studies as well as *in vivo* studies. To date, more frequently IP-10\CXCL10, MCP-1\CCL2, MIP-1alpha\CCL3, RANTES\CCL5, and IL-8\CXCL8 chemokines have been analyzed in periodontal disease and it was detected that their levels are significantly increased in inflamed tissues and decreased after non-surgical periodontal therapy. Thus, it was suggested that these chemokines play an important role in the pathogenesis of periodontal diseases and can be used as useful diagnostic biomarkers in the diagnosis of periodontal diseases. Moreover, with the analysis of chemokines in biological fluids, possible periodontal disease status can be revealed and irreversible periodontal tissue destruction can be prevented. However, the role of many different chemokines in periodontal disease has not been fully elucidated. Future studies need to explain different chemokines and cell-chemokine-receptor interactions in cellular and molecular events in periodontal disease.

Conflict of interest

The authors declare no conflict of interest.

Author details

Figen Öngöz Dede[1]* and Şeyma Bozkurt Doğan[2]

1 Faculty of Dentistry, Department of Periodontology, Ordu University, Ordu, Turkey

2 Faculty of Dentistry, Department of Periodontology, Ankara Yıldırım Beyazıt University, Ankara, Turkey

*Address all correspondence to: figen_ongoz@hotmail.com

References

[1] Rath-Deschner B, Memmert S, Damanaki A, de Molon RS, Nokhbehsaim M, Eick S, et al. CXCL5, CXCL8, and CXCL10 regulation by bacteria and mechanical forces in periodontium. Annals of Anatomy. 2021;**234**:151648. DOI: 10.1016/j.aanat.2020.151648

[2] Dezerega A, Pozo P, Hernández M, Oyarzún A, Rivera O, Dutzan N, et al. Chemokine monocyte chemoattractant protein-3 in progressive periodontal lesions in patients with chronic periodontitis. Journal of Periodontology. 2010;**81**(2):267-276. DOI: 10.1902/jop.2009.090406

[3] Hosokawa Y, Hosokawa I, Ozaki K, Nakae H, Matsuoet T. CXC chemokine ligand 16 in periodontal diseases: Expression in diseased tissues and production by cytokine-stimulated human gingival fibroblasts. Clinical and Experimental Immunology. 2007;**149**(1):146-154. DOI: 10.1111/j.1365-2249.2007.03398.x

[4] Haytural O, Yaman D, Ural EC, Kantarci A, Demirel K. Impact of periodontitis on chemokines in smokers. Clinical Oral Investigations. 2015;**19**(5):979-986. DOI: 10.1007/s00784-014-1314-2

[5] Gemmell E, Carter CL, Seymour GJ. Chemokines in human periodontal disease tissues. Clinical and Experimental Immunology. 2001;**125**(1):134-141. DOI: 10.1046/j.1365-2249.2001.01511.x

[6] Souto GR, Queiroz CM, Costa FO, Mesquita RA. Relationship between chemokines and dendritic cells in human chronic periodontitis. Journal of Periodontology. 2014;**85**(10):1416-1423. DOI: 10.1902/jop.2014.130662

[7] Stadler AF, Angst PDM, Arce RM, Gomes SC, Oppermann RV, Susin C. Gingival crevicular fluid levels of cytokines/chemokines in chronic periodontitis: A meta-analysis. Journal of Clinical Periodontology. 2016;**43**(9):727-745. DOI: 10.1111/jcpe.12557

[8] Havens AM, Chiu E, Taba M, Wang J, Shiozawa Y, Jung Y, et al. Stromal-Derived Factor-1α (CXCL12) levels increase in periodontal disease. Journal of Periodontology. 2008;**79**(5):845-853. DOI: 10.1902/jop.2008.070514

[9] Aldahlawi S, Youssef AR, Shahabuddin S. Evaluation of chemokine CXCL10 in human gingival crevicular fluid, saliva, and serum as periodontitis biomarker. Journal of Inflammation Research. 2018;**31**(11):389-396. DOI: 10.2147/JIR.S177188

[10] Emingil G, Atilla G, Hüseyinov A. Gingival crevicular fluid monocyte chemoattractant protein-1 and RANTES levels in patients with generalized aggressive periodontitis. Journal of Clinical Periodontology. 2004;**31**(10):829-834. DOI: 10.1111/j.1600-051X.2004.00584.x

[11] Pradeep AR, Daisy H, Hadge P. Gingival crevicular fluid levels of monocyte chemoattractant protein-1 in periodontal health and disease. Archives of Oral Biology. 2009;**54**(5):503-509. DOI: 10.1016/j.archoralbio.2009.02.007

[12] Pradeep AR, Daisy H, Hadge P. Serum levels of monocyte chemoattractant protein-1 in periodontal health and disease. Cytokine. 2009;**47**(2):77-81. DOI: 10.1016/j.cyto.2009.05.012

[13] Kawamoto D, Amado PPL, Albuquerque-Souza E, Bueno MR, Vale GC, Saraiva L, et al. Chemokines and cytokines profile in whole saliva of patients with periodontitis. Cytokine. 2020;**135**:155197. DOI: 10.1016/j.cyto.2020.155197

[14] Cardoso EM, Reis C. Chronic periodontitis, inflammatory cytokines, and interrelationship with other chronic diseases. Postgraduate Medicine. 2018;**130**(1):98-104. DOI: 10.1080/00325481.2018.1396876

[15] Slots J. Periodontitis: Facts, fallacies and the future. Periodontology. 2000;**75**(1):7-23

[16] Nagashima H, Shinoda M, Honda K, Kamio N, Watanabe M, Suzuki T, et al. CXCR4 signaling in macrophages contributes to periodontal mechanical hypersensitivity in Porphyromonas gingivalis -induced periodontitis in mice. Molecular Pain. 2017;**13**:1744. DOI: 10.1177/1744806916689269

[17] Tatakis DN, Trombelli L. Modulation of clinical expression of plaque-induced gingivitis: I. Background review and rationale. Journal of Clinical Periodontology. 2004;**31**(4):229-238. DOI: 10.1111/j.1600-051x.2004.00477.x

[18] Tatakis DN, Kumar PS. Etiology and pathogenesis of periodontal diseases. Dental Clinics of North America. 2005;**49**(3):491-516. DOI: 10.1016/j.cden.2005.03.001

[19] Caton J, Armitage G, Berglundh T, ILC C, Jepsen S, Kornman KS, et al. A new classification scheme for periodontal and peri-implant diseases and conditions – Introduction and key changes from the 1999 classification. Journal of Clinical Periodontology. 2018;**45**(20):S1-S8

[20] Armitage GC. Development of a classification system for periodontal diseases and conditions. Annals of Periodontology. 1999;**4**(1):1-6. DOI: 10.1902/annals.1999.4.1.1

[21] Kinane DF, Stathopoulou PG, Papapanou PN. Periodontal diseases. Nature Reviews. 2017;**22**(3):17038

[22] Papapanou PN, Sanz M, Buduneli N, Dietrich T, Feres M, Fine DH, et al. Periodontitis: Consensus report of workgroup 2 of the 2017 World Workshop on the Classification of Periodontal and Peri-Implant Diseases and Conditions. Journal of Periodontology. 2018;**89**:S173

[23] Tonetti MS, Greenwell H, Kornman KS. Staging and grading of periodontitis: Framework and proposal of a new classification and case definition. Journal of Periodontology. 2018;**89**:159-172. DOI: 10.1002/JPER.18-0006

[24] Zhu N, Zheng X, Qiao W, Huang W, Li R, Song Y. Activation of GATA-binding protein 4 regulates monocyte chemoattractant protein-1 and chemotaxis in periodontal ligament cells. Journal of Periodontal Research. 2022;**57**(1):195-204. DOI: 10.1111/jre.12953

[25] Zhuang Z, Yoshizawa-Smith S, Glowacki A, Maltos K, Pacheco C, Shehabeldin M, et al. Induction of M2 macrophages prevents bone loss in murine periodontitis models. Journal of Dental Research. 2019;**98**(2):200-208. DOI: 10.1177/0022034518805984

[26] Thunell DH, Tymkiw KD, Johnson GK, Joly S, Burnell KK, Cavanaugh JE, et al. A multiplex immunoassay demonstrates reductions in gingival crevicular fluid cytokines following initial periodontal therapy. Journal of Periodontal Research. 2010;**45**(1):148-152. DOI: 10.1111/j.1600-0765.2009.01204.x

[27] Behm C, Blufstein A, Gahn J, Noroozkhan N, Moritz A, Rausch-Fan X, et al. Soluble CD14 enhances the response of periodontal ligament stem cells to toll-like receptor 2 agonists. Mediators Inflammation. 2019;**23**:812

[28] Silva TA, Garlet GP, Fukada SY, Silva JS, Cunha FQ. Chemokines in oral inflammatory diseases: Apical periodontitis and periodontal disease. Journal of Dental Research. 2007;**86**(4):306-319. DOI: 10.1177/154405910708600403

[29] Kinane DF. Causation and pathogenesis of periodontal disease. Periodontology. 2000;**25**:8-20

[30] Repeke CE, Ferreira SB, Claudino M, Silveira EM, de Assis GF, Avila-Campos MJ, et al. Evidences of the cooperative role of the chemokines CCL3, CCL4 and CCL5 and its receptors CCR1+ and CCR5+ in RANKL+ cell migration throughout experimental periodontitis in mice. Bone. 2010;**46**(4):1122-1130. DOI: 10.1016/j.bone.2009.12.030

[31] Fredriksson MI. Effect of priming in subpopulations of peripheral neutrophils from patients with chronic periodontitis. Journal of Periodontology. 2012;**83**(9):1192-1199. DOI: 10.1902/jop.2012.110584

[32] Lee JS, Lee J. Chemokine in inflamed periodontal tissues activates healthy periodontal-ligament stem cell migration. Journal of Clinical Periodontology. 2017;**44**(5):530-539. DOI: 10.1111/jcpe.12710

[33] Kabashima H, Yoneda M, Nagata K, Hirofuji T, Maedaet K. The presence of chemokine (MCP-1, MIP-1α, MIP-1β, IP-10, RANTES)-positive cells and chemokine receptor (CCR5, CXCR3)-positive cells in inflamed human gingival tissues. Cytokine. 2021;**20**(2):70-77

[34] Shimada Y, Tabeta K, Sugita N, Yoshie H. Profiling biomarkers in gingival crevicular fluid using multiplex bead immunoassay. Archives of Oral Biology. 2013;**58**(6):724-730. DOI: 10.1016/j.archoralbio.2012.11.012

[35] Garlet GP, Martins W, Ferreira BR, Milanezi CM, Silva JS. Patterns of chemokines and chemokine receptors expression in different forms of human periodontal disease. Journal of Periodontal Research. 2003;**38**(2):210-217. DOI: 10.1034/j.1600-0765.2003.02012.x

[36] Moelants EAV, Loozen G, Mortier A, Martens E, Opdenakker G, Mizgalska D, et al. Citrullination and proteolytic processing of chemokines by Porphyromonas gingivalis. Infection and Immunity. 2014;**82**(6):2511-2519. DOI: 10.1128/IAI.01624-14

[37] Sakai A, Ohshima M, Sugano N, Otsuka K, Ito K. Profiling the cytokines in gingival crevicular fluid using a cytokine antibody array. Journal of Periodontology. 2006;**77**(5):856-864. DOI: 10.1902/jop.2006.050340

[38] Peyyala R, Kirakodu SS, Novak KF, Ebersole JL. Oral microbial biofilm stimulation of epithelial cell responses. Cytokine. 2012;**58**(1):65-72. DOI: 10.1016/j.cyto.2011.12.016

[39] Hosokawa Y, Hosokawa I, Shindo S, Ozaki K, Matsuo T. Calcitriol suppressed inflammatory reactions in IL-1β-stimulated human periodontal ligament cells. Inflammation. 2015;**38**(6):2252-2258. DOI: 10.1007/s10753-015-0209-y

[40] Panezai J, Ghaffar A, Altamash M, Sundqvist KG, Engström PE, Larsson A. Correlation of serum cytokines, chemokines, growth factors and enzymes with periodontal disease parameters. PLoS One. 2017;**12**(11):e0188945. DOI: 10.1371/journal.pone.0188945

[41] Nishikawa Y, Kajiura Y, Lew JH, Kido JI, Nagata T, Naruishi K. Calprotectin induces IL-6 and MCP-1 production via toll-like receptor 4 signaling in human gingival fibroblasts. Journal of Cellular Physiology. 2017;**232**(7):1862-1871. DOI: 10.1002/jcp.25724

[42] Yu X, Antoniades HN, Graves DT. Expression of monocyte chemoattractant protein 1 in human inflamed gingival tissues. Infection and Immunity. 1993;**61**(11):4622-4628. DOI: 10.1128/iai.61.11.4622-4628.1993

[43] Hanazawa S, Kawata Y, Takeshita A, Kumada H, Okithu M, Tanaka S, et al. Expression of monocyte chemoattractant protein 1 (MCP-1) in adult periodontal disease: Increased monocyte chemotactic activity in crevicular fluids and induction of MCP-1 expression in gingival tissues. Infection and Immunity. 1993;**61**(12):5219-5224. DOI: 10.1128/iai.61.12.5219-5224.1993

[44] Tonetti MS, Imboden MA, Gerber L, Lang NP, Laissue J, Mueller C. Localized expression of mRNA for phagocyte-specific chemotactic cytokines in human periodontal infections. Infection and Immunity. 1994;**62**(9):4005-4014. DOI: 10.1128/iai.62.9.4005-4014.1994

[45] Nebel D, Jönsson D, Norderyd O, Bratthall G, Nilssonet BO. Differential regulation of chemokine expression by estrogen in human periodontal ligament cells. Journal of Periodontal Research. 2010;**45**(6):796-802. DOI: 10.1111/j.1600-0765.2010.01308.x

[46] Jiang Y, Graves DT. Periodontal pathogens stimulate CC-chemokine production by mononuclear and bone-derived cells. Journal of Periodontology. 1999;**70**(12):1472-1478. DOI: 10.1902/jop.1999.70.12.1472

[47] Hanioka T, Takaya K, Matsumori Y, Matsuse R, Shizukuishi S. Relationship of the substance P to indicators of host response in human Gingival Crevicular Fluid. Journal of Clinical Periodontology. 2000;**27**(4):262-266. DOI: 10.1034/j.1600-051x.2000.027004262.x

[48] Bamashmous S, Kotsakis GA, Kerns KA, Leroux BG, Zenobia C, Chen D, et al. Human variation in gingival inflammation. Proceedings of the National Academy Science U S A. 2021;**118**(27):e2012

[49] Gupta M, Chaturvedi R, Jain A. Role of monocyte chemoattractant protein-1 (MCP-1) as an immune-diagnostic biomarker in the pathogenesis of chronic periodontal disease. Cytokine. 2013;**61**(3):892-897. DOI: 10.1016/j.cyto.2012.12.012

[50] Pradeep AR, Daisy H, Hadge P, Garg G, Thoratet M. Correlation of gingival crevicular fluid interleukin-18 and monocyte chemoattractant protein-1 levels in periodontal health and disease. Journal of Periodontology. 2009;**80**(9):1454-1461. DOI: 10.1902/jop.2009.090117

[51] Lee E, Yang YH, Ho YP, Ho KY, Tsai CC. Potential role of vascular endothelial growth factor, interleukin-8 and monocyte chemoattractant protein-1 in periodontal diseases. The Kaohsiung Journal of Medical Sciences. 2003;**19**(8):406-415. DOI: 10.1016/s1607-551x(09)70484-x

[52] Silva N, Dutzan N, Hernandez M, Dezerega A, Rivera O, Aguillon JC, et al. Characterization of progressive periodontal lesions in chronic periodontitis patients: Levels of chemokines, cytokines, matrix metalloproteinase-13, periodontal pathogens and inflammatory cells. Journal of Clinical Periodontology. 2008;**35**(3):206-214. DOI: 10.1111/j.1600-051X.2007.01190.x

[53] Souto GR, Queiroz-Junior CM, Costa FO, Mesquita RA. Smoking effect on chemokines of the human chronic periodontitis. Immunobiology. 2014;**219**(8):633-636. DOI: 10.1016/j.imbio.2014.03.014

[54] Tymkiw KD, Thunell DH, Johnson GK, Joly S, Burnell KK, Cavanaugh JE, et al. Influence of smoking on gingival crevicular fluid cytokines in severe chronic periodontitis. Journal of Clinical Periodontology. 2011;**38**(3):219-228. DOI: 10.1111/j.1600-051X.2010.01684.x

[55] Martins ES, César-Neto JB, Albuquerque-Souza E, Rebeis ES, Holzhausen M, Pannuti CM, et al. One-year follow-up of the immune profile in serum and selected sites of generalized and localized aggressive periodontitis. Cytokine. 2019;**116**:27-37. DOI: 10.1016/j.cyto.2018.12.019

[56] Shaddox LM, Wiedey J, Calderon NL, Magnusson I, Bimstein E, Bidwell JA, et al. Local inflammatory markers and systemic endotoxin in aggressive periodontitis. Journal of Dental Research. 2011;**90**(9):1140-1144. DOI: 10.1177/0022034511413928

[57] Kurtiş B, Tüter G, Serdar M, Akdemir P, Uygur C, Firatli E, et al. Gingival crevicular fluid levels of monocyte chemoattractant protein-1 and tumor necrosis factor-alpha in patients with chronic and aggressive periodontitis. Journal of Periodontology. 2005;**76**(11):1849-1855. DOI: 10.1902/jop.2005.76.11.1849

[58] Al-Sabbagh M, Alladah A, Lin Y, Kryscio RJ, Thomas MV, Ebersole JL, et al. Bone remodeling-associated salivary biomarker MIP-1α distinguishes periodontal disease from health. Journal of Periodontal Research. 2012;**47**(3):389-395. DOI: 10.1111/j.1600-0765.2011.01445.x

[59] Fine DH, Markowitz K, Furgang D, Fairlie K, Ferrandiz J, Nasri C. Macrophage inflammatory protein-1α: A salivary biomarker of bone loss in a longitudinal cohort study of children at risk for aggressive periodontal disease? Journal of Periodontology. 2009;**80**(1):106-113. DOI: 10.1902/jop.2009.080296

[60] Emingil G, Gürkan A, Atilla G, Kantarci A. Subantimicrobial-dose doxycycline and cytokine-chemokine levels in gingival crevicular fluid. Journal of Periodontology. 2011;**82**(3):452-461

[61] Gamonal J, Bascones A, Jorge O, Silva A. Chemokine RANTES in gingival crevicular fluid of adult patients with periodontitis. Journal of Clinical Periodontology. 2000;**27**(9):675-681. DOI: 10.1034/j.1600-051x.2000.027009675.x

[62] Gamonal J, Acevedo A, Bascones A, Jorge O, Silva A. Levels of interleukin-1 beta, -8, and -10 and RANTES in gingival crevicular fluid and cell populations in adult periodontitis patients and the effect of periodontal treatment. Journal of Periodontology. 2000;**71**(10):1535-1545. DOI: 10.1902/jop.2000.71.10.1535

[63] Gamonal J, Acevedo A, Bascones A, Jorge O, Silva A. Characterization of cellular infiltrate, detection of chemokine receptor CCR5 and interleukin-8 and RANTES chemokines in adult periodontitis. Journal of Periodontal Research. 2001;**36**(3):194-203. DOI: 10.1034/j.1600-0765.2001.360309.x

[64] Fokkema SJ, Loos BG, Van Der Velden U. Monocyte-derived RANTES is intrinsically elevated in periodontal disease while MCP-1

levels are related to inflammation and are inversely correlated with IL-12 levels. Clinical and Experimental Immunology. 2003;**131**(3):477-483. DOI: 10.1046/j.1365-2249.2003.02070.x

[65] Linhartova PB, Kavrikova D, Tomandlova M, Poskerova H, Rehka V, Dušek L, et al. Differences in interleukin-8 plasma levels between diabetic patients and healthy individuals independently on their periodontal status. International Journal of Molecular Science. 2018;**18**(10):3214

[66] Deinzer R, Weik U, Kolb-Bachofen V, Herforth A. Comparison of experimental gingivitis with persistent gingivitis: Differences in clinical parameters and cytokine concentrations. Journal of Periodontal Research. 2007;**42**(4):318-324. DOI: 10.1111/j.1600-0765.2006.00951.x

[67] Lappin DF, Robertson D, Hodge P, Treagus D, Awang RA, Ramage G, et al. The influence of glycated hemoglobin on the cross susceptibility between Type 1 diabetes mellitus and periodontal disease. Journal of Periodontology. 2015;**86**(11):1249-1259. DOI: 10.1902/jop.2015.150149

[68] Mesia R, Gholami F, Huang H, Clare-Salzler M, Aukhil I, Wallet SM, et al. Systemic inflammatory responses in patients with type 2 diabetes with chronic periodontitis. BMJ Open. 2016;**8**(1):e000260

[69] Mohamed HG, Idris SB, Ahmed MF, Åstrøm AN, Mustafa K, Ibrahim SO, et al. Influence of type 2 diabetes on local production of inflammatory molecules in adults with and without chronic periodontitis: A cross-sectional study. BMC Oral Health. 2015;**15**:86

[70] Engebretson SP, Vossughi F, Hey-Hadavi J, Emingil G, Grbicet JT. The influence of diabetes on gingival crevicular fluid β-glucuronidase and interleukin-8. Journal of Clinical Periodontology. 2006;**33**(11):784-790. DOI: 10.1111/j.1600-051X.2006.00984.x

[71] Longo PL, Artese HPC, Horliana ACRT, Gomes GH, Romito GA, Dib SA, et al. Inflammatory markers in gingival crevicular fluid of periodontitis patients with type 2 diabetes mellitus according to glycemic control: A pilot study. Dental Research Journal. 2015;**12**(5):449-455

[72] Tang X, Pan Y, Zhao Y. Vitamin D inhibits the expression of interleukin-8 in human periodontal ligament cells stimulated with Porphyromonas gingivalis. Archives of Oral Biology. 2013;**58**(4):397-407. DOI: 10.1016/j.archoralbio.2012.09.010

[73] AndrukhovO, AndrukhovaO, HulanU, Tang Y, Bantleon HP, Rausch-Fan X. Both 25-hydroxyvitamin-D3 and 1,25-dihydroxyvitamin- D3 reduces inflammatory response in human periodontal ligament cells. PLoS One. 2014;**28**(2):e90301

[74] Hosokawa Y, Hosokawa I, Ozaki K, Nakae H, Murakami K, Miyake Y, et al. CXCL12 and CXCR4 expression by human gingival fibroblasts in periodontal disease. Clinical and Experimental Immunology. 2005;**141**(3):467-474. DOI: 10.1111/j.1365-2249.2005.02852.x

[75] Lan KF, Shen YQ, Li Y, Ling CL, Gong YM, Xia SC, et al. Chemokine C-C motif ligand 8 in periodontal ligament during orthodontic tooth movement. Archives of Oral Biology. 2021;**123**:104996. DOI: 10.1016/j.archoralbio.2020.104996

[76] Oliveira SHP, Brito VGB, Frasnelli SCT, Ribeiro BS, Ferreira MN,

Queiroz DP, et al. Aliskiren attenuates the inflammatory response and wound healing process in diabetic mice with periodontal disease. Frontiers in Pharmacology. 2019;**4**(10):708

[77] Bonato CF. Hypertension favors the inflammatory process in rats with experimentally induced periodontitis. Journal of Periodontal Research. 2012;**47**(6):783-792. DOI: 10.1111/j.1600-0765.2012.01496.x

[78] Hernández Ríos M, Gamonal J, Salo T, Hukkanen M, Tjäderhane L, Sorsaet T. Reduced expression of lipopolysaccharide-induced CXC chemokine in Porphyromonas gingivalis-induced experimental periodontitis in matrix metalloproteinase-8 null mice. Journal of Periodontal Research. 2011;**46**(1):58-66. DOI: 10.1111/j.1600-0765.2010.01310.x

[79] Schallhorn RA, Patel DN, Chandrasekar B, Mealey BL. Periodontal disease in association with systemic levels of interleukin-18 and CXC ligand 16 in patients undergoing cardiac catheterization. Journal of Periodontology. 2010;**81**(8):1180-1186. DOI: 10.1902/jop.2010.100046

Chapter 5

Immunological Insights on Pathogenic Connections between Hepato-Digestive Disorders and Periodontal Conditions

Dora Maria Popescu, Dorin Nicolae Gheorghe, Flavia Nicolae, Lucian Paul Dragomir, Allma Pitru, Andrada Soanca, Cristina Florescu, Vasilica Toma, Liliana Georgeta Foia, Alexandra Livia Roman, Petra Surlin and Ion Rogoveanu

Abstract

The oral cavity is an integral part of the digestive tract and thus significant diseases, including periodontitis, can have an important impact on the normal nutritional functions of the body. Certain diseases of the hepato-digestive system have an inflammatory component, such as chronic hepatitis, fatty liver disease, or gastric cancer. This inflammatory reaction is mainly driven by pro-inflammatory chemokines. This is also the case for periodontitis, a condition characterized by the inflammation of the supporting tissues of teeth. Thus, significant pathogenic connections mediated by pro-inflammatory chemokines could exist between periodontitis and diseases of the hepato-digestive system.

Keywords: periodontitis, periodontal medicine, periodontal conditions, hepato-digestive disorders, pathogenic connections, immunology, inflammation mediators

1. Introduction

Low-grade inflammation is defined as the continuous, low-grade production of inflammatory factors throughout the body. It is currently acknowledged as a risk factor for many chronic diseases, including cancer, cardiovascular, cerebrovascular, and neurodegenerative illnesses [1, 2]. Periodontitis is a chronic inflammatory multifactorial disease, that involves the tooth-supporting structures, the periodontium, and in developed countries is the most frequent reason for tooth loss. Despite being essential, the infection is insufficient for the occurrence of this disease, as the dysregulation of the immuno-inflammatory status is required also, thus leading to systemic low-grade inflammation [1]. Periodontal microbes and immunity engage in a battle, with the help of innate immunity (such as macrophages, dendritic cells, natural killer cells, and neutrophils) and adaptive immunity (such as B and T lymphocytes), which leads to the release of pro-inflammatory molecules and enzymes (such as interferon-gamma, interleukin-17 (IL-17), tumor necrosis factor-α (TNF-α), interleukin-1 (IL-1), interleukin-6 (IL-6), collagenases such as matrix metalloproteinases (MMPs) [2].

On the one hand, our body uses the inflammatory response as a type of defense against microorganisms getting within the deeper tissues (such as bone). The periodontium, on the other hand, is irreversibly destroyed if the inflammation persists and is not controlled, leading to the classic symptoms of periodontitis, including periodontal pockets, attachment loss, gingival recessions, tooth mobility, tooth migration, and tooth loss [1, 2]. The interface via which local inflammation can have an impact on overall health is represented by the dento-gingival epithelial surface area, which includes every pocket epithelium in direct contact with the subgingival biofilm. Pro-inflammatory mediators produced locally during periodontitis, such as IL-1, IL-6, TNF-α, and prostaglandin E2 (PGE2), may enter the systemic circulation and then have an impact on distant organs, causing an inflammatory state to worsen and/or persist. Thus, the theory that the systemic inflammation brought on by periodontitis may influence the subject's inflammatory burden develops [1, 3].

Periodontal disease is considered a risk factor for a number of chronic illnesses, such as diabetes, cardiovascular disease, neurodegenerative disorders, or several malignancies, where low-grade inflammation is mandatory for the development and progression of the disease [1]. The mechanisms that associate periodontitis to extra-oral comorbidities are in line with clinical observations that connect periodontitis to bacteriemias, low-grade systemic inflammation, increased myelopoiesis, and the ability of local periodontal therapy to attenuate systemic inflammatory markers and improve comorbid disease activity [1–3]. Patients with severe periodontitis have increased levels of pro-inflammatory mediators (including IL-1, IL-6, C-reactive protein (CRP), and fibrinogen), as well as increased neutrophil counts in the blood, in comparison with healthy controls [3]. This could be because of oral bacteria, mainly *Porphyromonas gingivalis* and *Fusobacterium nucleatum*, which are partially responsible for periodontal disease and lead to a chronic inflammatory process, by damaging the fibroblasts, epithelial and endothelial cells, as they promote the release of inflammatory mediators, such as interleukin-1β (IL-1β), IL-6, IL-17, interleukin-23 (IL-23), TNF-α, and matrix metalloproteinases (MMP-8 and MMP-9) [2, 4].

Monocytes/macrophages, neutrophils, fibroblasts, and mast cells are the principal producers of IL-1β in the periodontium tissues in response to activation caused by lipopolysaccharide (LPS), the fundamental component of Gram-negative bacteria cell walls. Osteoclast development and bone resorption are brought on by IL-1β, which results in localized inflammation in the periodontium. Additionally, this cytokine induces the production of numerous MMPs, phospholipase A2, prostaglandins (PG), acute phase proteins, proinflammatory cytokine IL-6, and TNF-α. Regarding systemic malignancies, higher levels of this citokine are linked to tumor invasiveness, migration, and a more aggressive tumor phenotype. IL-1β creates an inflammatory milieu for angiogenesis and tumor growth [2, 4, 5].

IL-6 is a major pro-inflammatory cytokine. Numerous periodontal tissue cells produce it in response to stimulation when LPS and the pro-inflammatory cytokines IL-1 and TNF-α are present. In addition to causing bone resorption, IL-6 also promotes the production of acute phase proteins, chemokines, and PGE2. By boosting the production of matrix MMPs, IL-6 also influences the invasion and metastatic process [2, 4–6].

TNF-α, which is produced, among others, by monocytes/macrophages, neutrophils, fibroblasts, lymphocytes, and mast cells, is another important cytokine in the inflammatory response. In response to a variety of triggers, such as bacterial LPS, this cytokine is released. TNF-α significantly stimulates the formation of reactive oxygen species (ROS), leukotrienes, prostaglandins, and metalloproteinases and decreases the number of fibroblasts and osteogenic cells. Low dosages of this molecule are linked to tumor development as opposed to large concentrations of TNF-α, which are linked to tumor destruction. By stimulating the expression of MMPs and stimulating the production of several angiogenic factors, including interleukin-8 (IL-8) and basic fibroblast growth factor, TNF-α has been demonstrated to affect processes of motility and invasion [4, 5, 7]. A tumor phenotype can be induced by modest, persistent TNF-α production levels. The formation of reactive oxygen species and reactive nitrogen species (RNS), which may cause DNA damage and hence promote carcinogenesis, is the basis of a TNF-α tumor promotion mechanism. Cancer and TNF-α mediated inflammation are related [6, 7].

This chapter aims to exhibit the possible connections that may exist between periodontitis and some hepato-digestive disorders, mediated by means of pro-inflammatory chemokines. These diseases share an inflammatory component as part of their pathogenic mechanisms, and thus, mutual pro-inflammatory chemokines could play a significant influence on their onset and evolution, in terms of susceptibility and severity.

2. Cytokines' involvement in periodontitis-systemic conditions

2.1 Periodontitis and colorectal cancer

A prevalent inflammatory illness with infectious origins, periodontitis frequently turns into a chronic condition. In addition to its significance as a stomatological condition, chronic periodontitis has gained significance because it has been demonstrated that it can progress to a systemic condition marked by unresolved hyper-inflammation, disruption of the innate and adaptive immune system,

dysbiosis of the oral and gut mucosa, and other conditions that may result in, coexist with or exacerbate other health problems while it is linked to increased morbidity and mortality. It is still up for discussion how the location's microbiota and other system-wide changes connect to the infectious, immunological, inflammatory, and systemic characteristics of periodontitis and gastrointestinal disorders [1, 2].

Similar to other diseases that we shall address later, systemic inflammation is considered to have an important biological function in these conditions as well. It is generally known that inflammation may have a significant impact on all stages of cancer. From the first cancerous cell to the first stages of tumor development, progression, and neoplasm spread, inflammatory and immunological mediation processes are well-known markers of cancer. Another developing idea in this context is that, similar to other complicated illnesses, cancer develops from systemic rather than local factors. A complicated process by itself, systemic inflammation involves interactions between immunological signals, energy metabolism, and functional linkages that, when combined with genetic instability, predispose people to cancer and control the aberrant conditions that support neoplastic illness [1–3].

In the western world, colorectal cancer (CRC) is the second most common cause of cancer-related mortality. The several stages of CRC formation are defined by complicated interactions between environmental carcinogens, genetic changes, and the host immune system, which eventually lead to the uncontrolled expansion of altered cells. Chronic inflammation is a distinct risk factor for the development of CRC, just like it is for other prevalent malignancies (including gastric cancer, prostate cancer, and hepatocellular carcinoma). According to experimental models of inflammation-related colon carcinogenesis, cytokines produced by inflammatory cells can either directly or indirectly drive the development of cancer cells [8]. Moreover, a higher risk of colorectal cancer has been linked to plasma levels of certain pro-inflammatory cytokines, such as IL-8 and IL-6 [9].

In a research paper, tumor cells were shown to be a significant source of chemokines in CRC, and the gut microbiota was found to have a key role in chemokine synthesis and T cell recruitment in tumor tissue, both of which improved prognosis. This information may potentially pave the path for the creation of novel therapies focused on altering gut flora to encourage immune cell populations with positive prognostic relevance to infiltrate CRCs [10]. Chemokines primarily control angiogenesis, activate immune responses specific to tumors, and stimulate the tumor directly through autocrine or paracrine processes in cancer [11].

F. nucleatum, an periopathogen, can hematogenously spread to the colon, where it prefers CRC, as it directly binds to and invades host immune and cancer cells [3, 11]. This is because D-galactose-N-Acetyl-D-galactosamine (GAL-GalNAc), a carbohydrate moiety that the *F. nucleatum* lectin Fap2 binds, is overexpressed in CRC cells. This overexpression also triggers the release of prometastatic chemokines like IL-8 and chemokine family CXC-chemokine ligand 1 (CXCL1), which promote CRC migration and tumor-related angiogenesis. By activating E-cadherin-mediated Wnt—catenin signaling in a way that is reliant on annexin A1, which is increased in CRC, another *F. nucleatum* adhesin called FadA promotes the proliferation of CRC (but not non-cancerous) cells [3, 4].

For instance, when 13 cytokines, chemokines, and growth factors were examined in serum profiles from 116 CRC patients and 86 healthy controls, it was discovered that five of these proteins had statistically significant changes in their serum levels, including elevated levels of IL-6, IL-7, CXCL-8 (IL-8), and platelet-derived growth factor-subunit B (PDGFB) and decreased levels of chemokine ligand 2 (CCL2) [12]. Also, pro-inflammatory cytokines, such as IL-6, TNF-α, and IL-8, were found to be expressed more often in CRC patients than in the control group, and also more advanced CRC stages (stage III/IV vs. stage I/II cancers) are typically related with patients who have greater TNF-α concentrations [5]. In comparison to untreated mice, anti-TNF-treated animals showed considerably fewer colon tumors and lower histology scores of colon inflammation [13] while another study reported that the TNF-α serum level of the colorectal cancer group was statistically significantly lower than that of the control group [14].

Notably, IL-6 is one of the cytokines that are significantly higher in CRC patients compared to healthy controls and is much higher in metastatic cancer compared to non-metastatic disease, as it was reported that it causes CRC cells to proliferate, invade and migrate [7, 12, 15]. In comparison to IL-6 deletion tumors, those with IL-6 overexpression tended to develop more quickly, since angiogenesis is stimulated. Patients with increased IL-6 expression in CRC tissues had shorter overall survival (25.5 months on average) than those with lower IL-6 expression (46 months), as it was reported that IL-6 promotes chemotherapy resistance [5, 16]. When pro-inflammatory factors (IL-6, IL-1β, and TNF-α) were measured in the blood of patients with advanced cancer, IL-1β levels linked more significantly with clinical features than

IL-6 levels did [5]. Another research paper stated that, in CRC patients, IL-6 levels often correlate with tumor size, stage, and metastasis. In numerous kinds of cancer, circulating IL-6 levels are predictors of survival and response to therapy [16]. A postoperatively higher carcinoembryonic antigen (CEA) alone or in conjunction with carbohydrate antigen19–9 (CA19–9), chitinase-3-like-protein-1 (CHI3L1), CRP, or IL-6 may also signal individuals at high risk of relapsing, for CRC patients in stage II to IV who underwent radical surgery and received adjuvant 5-fluorouracil (5-FU)-based treatment [17].

2.2 Periodontitis and nonalcoholic fatty liver disease

The synthesis of inflammatory mediators is a characteristic of the systemic inflammatory response in periodontal disease, which is represented by periodontal infection associated with oral biofilm [18]. Even in the absence of other chronic systemic disorders, patients with chronic periodontitis have higher levels of proinflammatory mediators [19, 20]. It can be a bidirectional relationship between periodontitis and the majority of common noncommunicable diseases, including cardiovascular conditions, metabolic syndromes, diabetes mellitus and non-alcoholic fatty liver disease, cancer, and respiratory conditions [21].

Non-alcoholic fatty liver disease (NAFLD), is the most prevalent type of chronic liver disease globally. NAFLD is generally asymptomatic, and it is closely associated with diseases like obesity, diabetes mellitus type 2, and the characteristics of metabolic syndrome [22]. Scientific evidence strongly indicates that the gut microbiota contributes significantly to the pathogenesis of NAFLD and that *P. gingivalis* (P. g.) changes the composition of the intestinal microflora [23]. According to studies, P. g. infection is associated with various systemic disorders, such as diabetes mellitus, hepatic conditions, and gastric cancers [24]. Periopathogen bacteria such as P. g. can be detected, in association with a statistically significant decrease in serum albumin levels in NAFLD patients, indicating liver function impairment. Some studies concluded that P. g. infection could be an independent predictor of NAFLD development. These studies suggest that persistent P. g. infection in patients with untreated periodontitis may accelerate liver tissue fibrosis and decrease liver function [25].

A chronic low-grade inflammatory state is a pathological component of a variety of chronic conditions, including NAFLD [26]. Microvascular endothelial dysfunction, decreased immune responses, and chronic low-grade inflammation are all characteristics of the multisystem disease NAFLD [27]. A chronic low-grade inflammatory condition that is supported by the relationship between NAFLD, obesity, and type 2 diabetes increases the risk of atherosclerosis, dyslipidemia, systemic arterial hypertension, and acute myocardial infarction [28].

Simple steatosis and non-alcoholic steatohepatitis (NASH), which is marked by liver inflammation and hepatocyte ballooning with or without fibrosis, are also examples of NAFLD. NAFLD and NASH are linked to higher rates of morbidity and mortality across this broad spectrum of diseases [29].

Numerous variables, such as the patient's age and gender, ethnicity, frequency, amount of alcohol consumed daily, diet, hormonal status, genetic predisposition, microbiome, and metabolic status, have an impact on the heterogeneity in the clinic and the evolution of the fatty liver disease. There may be a differential impact on the contribution of these variables to each patient on the part or between patients, an impact that then changes the phenotype and evolution of the disease [30].

From in vitro, in vivo, and epidemiologic perspectives, the potential correlations between periodontitis and NAFLD have been studied, but the genetic and molecular processes of the relationship between periodontitis and NAFLD remain unknown [31]. According to epidemiological research, periodontitis and NAFLD together are increasing the occurrence of the condition, which could enhance the potential that it will advance to liver fibrosis as well [32]. However, alterations in the oral microbiome caused by periodontal disease could have an impact on the intestine microbiota and NAFLD's pathogenesis could be influenced by this gut microbiota [33].

The involvement of cytokines as important mediators of fibrosis, cirrhosis, and inflammation in NAFLD is generally accepted. Different inflammatory mediators, including IL-1, IL-6, TNF-α, CRP, and NOD-like receptor protein 3 (NLRP3) inflammasome, have been implicated in the pathogenesis and progression of NAFLD in previous studies [23, 34]. A systemic indicator of inflammation, CRP may be essential in detecting pathological alterations. The liver is the primary source of CRP but can also be secreted by adipose tissue, which acts as an endocrine organ that can secrete other inflammatory cytokines such as IL-6 [35]. A higher serum level of CRP has been identified in several studies as a significant risk factor for the development of NAFLD, and it appears to be a great marker for the prediction and diagnosis of NAFLD. A family of proteins known as pentraxin (PTX) is divided into two classes, short and long, depending on the length of their structural components. Serum amyloid P and CRP are two minor members of this family. One of the long proteins in this family, plasma pentraxin 3 (PTX3), is detected in significantly higher concentrations in patients with steatohepatitis than in those without [36]. Patients with more severe forms of NAFLD have higher serum concentrations of this protein, and higher levels of this protein are associated with more severe forms of hepatic fibrosis. Therefore, this could suggest that the serum PTX3 level might be used to diagnose the degree of hepatic fibrosis and to separate steatohepatitis and simple steatosis [37].

Particularly the pro-inflammatory cytokines IL-1 and TNF mediate important features of liver diseases including acute phase protein synthesis, lipid metabolism, cholestasis, and degree of fibrosis in different stages of liver diseases. These essential cytokines, which are mostly generated by mononuclear cells, influence all types of liver cells and control the generation of many other mediators essential in chronic liver diseases [38].

Other interleukins implicated in NAFLD disease include IL-8, IL-12, IL-18, and IL-34 [39]. Also, IL-32, which was first identified in 2005 and is one of several inflammatory biomarkers linked to obesity and NAFLD, is identified as a key regulator of obesity-driven inflammation and lipotoxicity [40].

According to some research, patients with NAFLD who ultimately progressed to critical condition had higher concentrations of IL-6, IL-8, and IL-10, than when establishing the diagnosis. IL-8 and IL-10 also seem to be significant prognostic biomarkers associated with recovery time [27]. IL-6 is often synthesized in immunologically activated adipose tissue, which is deeply associated with metabolic syndrome. This could be a possible explanation for the high prevalence of obesity in patients with NAFLD. Other studies did not find differences in IL-6 levels between non-obese and obese patients with NAFLD, suggesting other possible immunological mechanisms [27, 41].

The development of metabolic changes and an increased risk of liver damage associated with fibrosis during the evolution of NAFLD may be caused by inadequate modulation of IL-10 associated with IL-6, IL-12, and TNF [42]. Some proinflammatory cytokines (including IL-6, IL-12, and TNF-α) are considered to contribute to inadequate modulation of IL-10, according to some studies. Since IL-10 concentrations indicated a strong correlation with IL-6 and other inflammatory markers, including CRP, this could be interpreted as an unsuccessful attempt to reduce the hyperinflammatory response and tissue damage. Additionally, immune cells that have been stimulated may become "IL-10 resistant," which would allow them to pass the anti-inflammatory effects of IL-10 signaling and increase the inflammatory response [43]. Higher IL-6 levels and a lower IL-10/IL-6 ratio have been reported in NASH patients, which may indicate inadequate anti-inflammatory compensation. Patients with NASH have also been found to have higher levels of circulating TNF than patients with steatosis or healthy controls [42].

When NAFLD first appears and progresses, cytokines may contribute, stimulating and controlling critical processes. Patients with non-alcoholic steatohepatitis may have a balance between proinflammatory (IL-1, IL-6, IL-12, and TNF) and regulatory (IL-10) cytokine concentrations, which have been associated with early metabolic abnormalities in the context of NAFLD [42]. Furthermore, it has been shown that TNF production may have a role in the first liver injury, causing the release of other cytokines including IL-12, which recruits inflammatory cells, destroys hepatocytes, and initiates a healing response, including hepatic fibrogenesis [44]. Because of its antioxidant and cytoprotective characteristics, IL-12 may be a protective biomarker for NAFLD. Regardless of the number of metabolic risk factors, IL-12 was inversely correlated with the prevalence of NAFLD and positively correlated with the specific rise in productivity bilirubin in patients with non-alcoholic steatohepatitis [45].

2.3 Periodontitis and chronic hepatitis C

Chronic hepatitis C (CHC) is caused by infection with the Hepatitis C Virus (HCV) and is a severe threat to the individual's life, with an estimated 700,000 deaths worldwide each year [46]. The spread of the virus is considered a global health hazard since it affects more than 200 million individuals globally, and it is especially difficult to battle because the disease has no symptoms in its early stages. As a result, infected individuals may be oblivious and easily infect others. Most individuals will develop chronic liver inflammation after the acute stage of illness. Hepatic function gradually deteriorates when hepatic tissue is replaced by fibrotic tissue and liver cirrhosis develops [47].

CHC is a potentially fatal illness if left untreated, leading to serious consequences such as hepatic cirrhosis (in 10–30% of CHC patients) and hepatocellular cancer (5 percent of CHC patients) [48]. Only 15% of individuals have clinical manifestations of the illness after infected contact with the virus. The initial inflammatory reaction generated by the presence of the virus is "chronicized" in the majority of cases, resulting in CHC, with the virus remaining detectable six months after the infected encounter [49].

Patients with CHC frequently experience significant oral health issues, which can have a negative impact on their quality of life, in addition to the pathological manifestations of liver disease and its complications [50]. CHC patients seeking dental treatments may encounter a variety of difficulties, such as high personal anxiety or altered healing and recovery processes following dental and periodontal surgery, which restrict the complexity of therapy alternatives [51]. CHC patients may have increased risk factors for the start of periodontitis, leading to its clinical presentation, which is driven by the buildup of subgingival bacterial plaque deposits, which is corroborated by probable behavioral differences [52].

As a chronic inflammatory disorder, CHC may have certain connections and interactions with Parkinson's disease, presumably through proinflammatory mediators released into the bloodstream of HCV patients [53]. Natural killer cells (NK), which may produce TNF-α, play a crucial role in the immunopathogenesis of CHC [54]. Certain cytokines, including IL-18 and IL-33, are employed as indicators of CHC disease activity and severity since these individuals have higher serological levels of these interleukins [55, 56]. Elevated levels of IL-1α have also been detected in blood samples of chronic hepatitis C patients, and have been linked to disease severity [57]. Furthermore, several proinflammatory mediators, including IL-1α, have been demonstrated to stimulate hepatic inflammatory processes in chronic hepatitis C patients, resulting in elevated blood levels [58].

In essence, both CHC and periodontitis cause a persistent inflammatory response, with proinflammatory mediators controlling the amount and intensity of the pathogenic process. Studies that focused on the gingival crevicular fluid (GCF) assessment of IL-1α and IL-1β's involvement in the pathogenic process of periodontitis patients with CHC, found that patients with both diseases had significantly worsened periodontal status and higher levels of these cytokines than non-CHC patients with periodontitis [59]. This suggests that hepatic pathology may have a negative impact on local periodontal inflammation.

Elevated gingival fluid levels of IL-1α and IL-1β, detected in periodontal patients compared to healthy controls, have been demonstrated to decrease following periodontal therapy, supporting their critical involvement in the development of periodontal disease [24, 59]. A statistically significant difference in GCF cytokine levels between the CHC and periodontitis groups was also highlighted. Because both groups of patients had periodontal disease with equal degrees of severity and progression, the greater GCF cytokines levels in periodontitis patients with CHC might be explained by the added chronic hepatic inflammation that these individuals exhibit. This fact can also influence the severity of the inflammatory periodontal response [59].

The levels of GCF cytokines in chronic periodontitis patients correlate with the degree of periodontal inflammation. The clinical markers used to measure periodontal condition (such as the number of missing teeth, periodontal pocket depth, or gingival attachment loss) were associated to increased GCF levels of IL-1α and IL-1β in periodontal patients with CHC [59]. There was also a modest positive connection between these clinical measures in periodontal patients with no systemic disease and GCF cytokine levels, indicating that systemic chronic inflammation might have an extra influence on such patients' periodontal health [59].

Periodontal and hepatic chronic inflammatory responses may interact because they are both fueled by the same proinflammatory cytokines, IL-1α and IL-1β [60]. Hepatic chronic inflammation has a significant detrimental influence on periodontal condition in terms of the strength of the periodontal inflammatory reaction [61]. Elevated levels of IL-1α and IL-1β, which have significant consequences in the pathogenic processes of both periodontitis and CHC, might suggest that chronic hepatitis C has a deleterious influence on periodontal patients' inflammatory status, as measured by IL-1α and IL-1β GCF detection.

The notion of the "inflammasome" molecule has opened up new perspectives for periodontal inflammation research. The NLPR3 inflammasome is a critical component of this inflammatory response, serving as the first triggering mechanism upon interaction with bacterial antigens such as *P.g's* LPS [62]. Substantial variations in

GCF NLRP3 levels between the periodontitis and control groups were reported, as well as between patients with CHC and periodontitis and CHC-only ones, suggesting that periodontal pathological events cause a significant rise in NLRP3 expression [63]. This concept was also evaluated and supported by a recent study on the issue, which indicated that periodontal disorders are characterized by an increase in inflammasome expression and a decrease in their inhibitor proteins [64, 65]. In terms of hepatic disease, NLRP3 has been shown to activate inside white and red blood cells when triggered by HCV infection [66]. Furthermore, the *in vitro* presence of the virus dictated cellular pyroptosis within infected hepatocytes, an event primarily mediated by NLRP3 and caspase-1 (CASP1) activity [67, 68]. Together with IL-1, NLRP3 is implicated in the initiation of the chronic hepatic inflammatory response caused by HCV infection. As a result, CHC patients are likely to have higher blood NLRP3 levels [67].

According to research on the issue of NLRP3 inflammasome involved in the periodontitis and diabetes connection, the findings revealed substantial correlations between GCF NLRP3 levels and blood glucose levels of the participating patients [63, 69]. This result supports the pathogenic relationships that exist between periodontitis and insulin resistance. Similar pathogenic links exist between CHC and insulin resistance. These findings imply that this pathologic mechanism may be able to bridge the gap between chronic hepatitis and periodontitis due to the significant influence and bi-directional repercussions it has on the inflammatory response [69].

The immunological analysis produced equivalent results on the average values of CASP1 in periodontitis patients' GCF samples, which were considerably greater than those of healthy controls [65, 69]. CASP1 is largely expressed in gingival epithelial cells, keratinocytes, and connective tissue cells, and is essentially non-existent when periodontal tissues are not inflamed [70]. Furthermore, key periodontal bacteria (*Aggregatibacter actinomycetemcomitans* and *P. g.*) have been found to induce caspase expression in epithelial cells and macrophages [71]. CASP1 is also implicated in the pathology of CHC, since HCV-infected cells may synthesize and release the NLRP3 inflammasome [68]. The findings revealed that CHC + periodontitis patients had higher average GCF CASP1 levels than non-CHC periodontitis patients. There were also significant variations in the average values of CHC patients with and without periodontal disease. These findings imply that chronic inflammation, whether hepatic or periodontal, can have a considerable influence on CASP1 levels [65, 69].

IL-18 has also been proposed as a potential marker for periodontal structural degradation [72]. GCF IL-18 levels in periodontal patients were substantially greater than in healthy controls. Furthermore, the statistical analysis revealed substantial associations between GCF IL-18 levels and periodontal disease severity measures (periodontal probing depth, clinical attachment loss, and gingival bleeding index) [63, 69]. In line with this observation, CHC + periodontitis patients had substantially higher GCF IL-18 levels, indicating an unfavorable clinical periodontal state [64, 69]. Concerning hepatic illness, IL-18 has immunological and clinical repercussions for CHC, as affected individuals frequently have much higher levels of this mediator than healthy controls [73]. Significant variations in GCF IL-18 levels were seen between the CHC+ periodontitis patients and non-CHC periodontitis ones, confirming the deleterious influence that HCV infection can have on periodontal status in comparable periodontal pathological circumstances [69]. This can exacerbate gingival pathogenic events and cause a more severe periodontal inflammatory response.

PTX is a protein family that participates in inflammatory pathways by serving as pattern-recognition receptors (PRRs) [74]. PTX, also known as acute-phase proteins (APP), play a key role in the initiation of inflammation, primarily by protecting against pathogenic bacteria via complement activation. This role is supported by the fact that in GCF samples originating from patients diagnosed with periodontitis, PTX3 levels were significantly elevated, as compared to control groups [36, 75]. These elevated GCF PTX3 levels also correlated with clinical periodontal parameters, suggesting an early up-regulated synthesis of this mediator by neutrophil cells, in response to periodontal bacterial aggression [36]. It is known that PTX3 is pre-stored in neutrophil cells to be accessible for quick release and action [75]. It was highlighted that GCF PTX3 levels rise proportionally to the severity of periodontitis. At baseline, GCF PTX3 levels were higher in periodontitis patients than in controls, and these levels considerably decreased as soon as two weeks following scaling and root planning (non-surgical periodontal therapy—NSPT) on these patients [76]. However, a higher fall in GCF PTX3 levels was found in smoking patients, indicating that smoking had a major influence on this parameter. GCF PTX3 levels in smoking patients remained increased even after NSPT [76].

CRP is a short pentraxin, whereas PTX3 belongs to the long pentraxin branch. Both proteins are increased in severe infections and inflammation, indicating their importance in bacterial or viral diseases (such as periodontitis and CHC) [77]. GCF CRP levels express lower association strength and, as a result, less periodontal clinical significance than the PTX3 mediator [36, 76]. GCF and CRP levels are considerably higher in periodontitis patients' samples compared to healthy controls [78]. However, there was no statistically significant difference in marker levels between periodontitis patients with and without CHC [36]. GCF CRP levels dropped after NSPT in CHC + periodontitis patients but increased somewhat in periodontitis-only patients. GCF CRP levels are associated with patients' age and the dental plaque index, indicating that it may be more important to senior patients, who have poorer dental hygiene. GCF CRP levels were also associated with the bleeding on the probing index [36]. These factors match the profile of CHC patients, who had much higher bleeding on probing than non-CHC individuals. This feature can be explained by the hepatic nature of CRP, which, unlike PTX3, is not locally pre-synthesized and may take a longer time to reach critical quantities in the GCF. This idea was supported by a study highlighting that, while NSPT improved periodontal clinical parameters such as plaque index (due to improved oral hygiene) and pocket depth/attachment loss (due to decreased inflammation), serum CRP levels did not decrease significantly after a one- and two-month recall [79]. Putative participation of hepatic disease in the local inflammatory processes of periodontitis necessitates more investigation in the future, similar to a larger perspective on the role of the CRP mediator in a shared periodontitis-CHC scenario.

3. Conclusions

Periodontitis is characterized by a low-grade inflammation, which triggers the local up-regulation of significant pro-inflammatory mediators, similar to that of certain systemic diseases (**Table 1**). Given that many hepato-digestive disorders have an important inflammatory component, the two types of conditions could exert a significant influence on each other, by means of mutual pro-inflammatory chemokines.

Inflammatory marker	Periodontitis/colorectal cancer	Periodontitis/ nonalcoholic fatty liver disease	Periodontitis/ chronic Hepatitis C
CASP1			*
CCL2	*		
CRP		*	*
CXCL1	*		
GAL-GalNAc	*		
IL-1		*	
IL-1α			*
IL-1β	*		*
IL-6	*	*	
IL-7	*		
IL-8	*	*	
IL-10		*	
IL-12		*	
IL-18		*	*
IL-32		*	
IL-33			*
IL-34		*	
NRLP3		*	*
PTX3		*	*
TNF-α	*	*	*

**-existing relevant information on implications in periodontitis and systemic diseases (CASP1 – caspase1; CCL2-chemokine ligand2; CRP – C Reactive Protein; CXCL1- CXC-chemokine ligand 1; GAL-GalNAc - D-galactose-N-Acetyl-D-galactosamine; IL – interleukin; NRLP3 - NOD-like receptor protein 3; PTX3 – pentraxin3; TNF-α – tumor necrosis factor-α.*

Table 1.
Synopsis of mediators' inflammation in the association of periodontitis and hepato-digestive disorders.

Conflict of interest

The authors declare no conflict of interest.

Author details

Dora Maria Popescu[1], Dorin Nicolae Gheorghe[1*], Flavia Nicolae[1], Lucian Paul Dragomir[2], Allma Pitru[3], Andrada Soanca[4], Cristina Florescu[5], Vasilica Toma[6], Liliana Georgeta Foia[7], Alexandra Livia Roman[4], Petra Surlin[1] and Ion Rogoveanu[8]

1 Faculty of Dental Medicine, Department of Periodontology, University of Medicine and Pharmacy of Craiova, Craiova, Romania

2 Faculty of Dental Medicine, Department of Fixed Prosthodontics, University of Medicine and Pharmacy of Craiova, Craiova, Romania

3 Faculty of Dental Medicine, Department of Oral Pathology, University of Medicine and Pharmacy of Craiova, Craiova, Romania

4 Faculty of Dental Medicine, Department of Periodontology, "Iuliu Hatieganu" University of Medicine and Pharmacy, Cluj-Napoca, Romania

5 Faculty of Medicine, Department of Internal Medicine and Cardiology, University of Medicine and Pharmacy of Craiova, Craiova, Romania

6 Faculty of Dental Medicine, Department of Pedodontics, "Gr. T. Popa" University of Medicine and Pharmacy, Iasi, Romania

7 Faculty of Dental Medicine, Department of Biochemistry, "Gr. T. Popa" University of Medicine and Pharmacy, Iasi, Romania

8 Faculty of Medicine, Department of Gastroenterology, University of Medicine and Pharmacy of Craiova, Craiova, Romania

*Address all correspondence to: dorinngheorghe@gmail.com

References

[1] Cecoro G, Annunziata M, Iuorio MT, Nastri L, Guida L. Periodontitis, low-grade inflammation and systemic health: A scoping review. Medicina (Kaunas, Lithuania). 2020;**56**(6):272

[2] Martínez-García M, Hernández-Lemus E. Periodontal inflammation and systemic diseases: An overview. Frontiers in Physiology. 2021;**12**:709438

[3] Hajishengallis G, Chavakis T. Local and systemic mechanisms linking periodontal disease and inflammatory comorbidities. Nature Reviews. Immunology. 2021;**21**(7):426-440

[4] Karpinski TM. Role of oral microbiota in cancer development. Microorganisms. 2019;**7**(1):20

[5] Kasprzak A. The role of tumor microenvironment cells in colorectal cancer (CRC) Cachexia. International Journal of Molecular Sciences. 2021;**22**(4):1565

[6] Landskron G, De la Fuente M, Thuwajit P, Thuwajit C, Hermoso MA. Chronic inflammation and cytokines in the tumor microenvironment. Journal of Immunology Research. 2014;**2014**:149185

[7] Kraus S, Arber N. Inflammation and colorectal cancer. Current Opinion in Pharmacology. 2009;**9**(4):405-410

[8] De Simone V, Franzè E, Ronchetti G, Colantoni A, Fantini MC, Di Fusco D, et al. Th17-type cytokines, IL-6 and TNF-α synergistically activate STAT3 and NF-kB to promote colorectal cancer cell growth. Oncogene. 2015;**34**(27):3493-3503

[9] Momen-Heravi F, Babic A, Tworoger SS, Zhang L, Wu K, Smith-Warner SA, et al. Periodontal disease, tooth loss and colorectal cancer risk: Results from the nurses' health study. International Journal of Cancer. 2017;**140**(3):646-652

[10] Cremonesi E, Governa V, Garzon JFG, Mele V, Amicarella F, Muraro MG, et al. Gut microbiota modulate T cell trafficking into human colorectal cancer. Gut. 2018;**67**(11):1984-1994

[11] Casasanta MA, Yoo CC, Udayasuryan B, Sanders BE, Umaña A, Zhang Y, et al. Fusobacterium nucleatum host-cell binding and invasion induces IL-8 and CXCL1 secretion that drives colorectal cancer cell migration. Science Signaling. 2020;**13**(641):eaba9157

[12] Tuomisto AE, Mäkinen MJ, Väyrynen JP. Systemic inflammation in colorectal cancer: Underlying factors, effects, and prognostic significance. World Journal of Gastroenterology. 2019;**25**(31):4383-4404

[13] Yang Y, Gharaibeh RZ, Newsome RC, Jobin C. Amending microbiota by targeting intestinal inflammation with TNF blockade attenuates development of colorectal cancer. Nature Cancer. 2020;**1**(7):723-734

[14] Coşkun Ö, Öztopuz Ö, Özkan ÖF. Determination of IL-6, TNF-α and VEGF levels in the serums of patients with colorectal cancer. Cellular and Molecular Biology (Noisy-le-Grand, France). 2017;**63**(5):97-101

[15] Zhang X, Hu F, Li G, Li G, Yang X, Liu L, et al. Human colorectal cancer-derived mesenchymal stem cells promote colorectal cancer progression through IL-6/JAK2/STAT3 signaling. Cell Death & Disease. 2018;**9**(2):25

[16] Hu F, Song D, Yan Y, Huang C, Shen C, Lan J, et al. IL-6 regulates autophagy and chemotherapy resistance by promoting BECN1 phosphorylation. Nature Communications. 2021;**12**(1):3651

[17] Hermunen K, Soveri LM, Boisen MK, Mustonen HK, Dehlendorff C, Haglund CH, et al. Postoperative serum CA19-9, YKL-40, CRP and IL-6 in combination with CEA as prognostic markers for recurrence and survival in colorectal cancer. Acta Oncologica. 2020;**59**(12):1416-1423

[18] Hatasa M, Yoshida S, Takahashi H, Tanaka K, Kubotsu Y, Ohsugi Y, et al. Relationship between NAFLD and periodontal disease from the view of clinical and basic research, and immunological response. International Journal of Molecular Sciences. 2021;**22**(7):3728

[19] Popescu DM, Pitru AR, Florescu C, Gheorghe DN, Ionele CM, Rica AM, et al. Implications of lipid and carbohydrate metabolism in periodontal disease associated with diabetes, cardiovascular and nonalcoholic fatty liver disease. Medical Surgical Journal. 2021;**125**(3):513-520

[20] Bartold PM, Van Dyke TE. Host modulation: Controlling the inflammation to control the infection. Periodontology 2000 2000. 2017;75(1):317-329

[21] Cardoso EM, Reis C, Manzanares-Céspedes MC. Chronic periodontitis, inflammatory cytokines, and interrelationship with other chronic diseases. Postgraduate Medicine. 2018;**130**(1):98-104

[22] Arsenie C, Săndulescu D, Popescu DM, Gheorghe DN, Mârţu A, Foia L, et al. Periodontal changes, and the non-alcoholic fatty liver disease. International Journal of Medical Dentistry. 2018;**22**(3):280-287

[23] Ponziani FR, Bhoori S, Castelli C, Putignani L, Rivoltini L, Del Chierico F, et al. Hepatocellular carcinoma is associated with gut microbiota profile and inflammation in nonalcoholic fatty liver disease. Hepatology. 2019;**69**(1):107-120

[24] Gheorghe DN, Camen A, Popescu DM, Sincar C, Pitru A, Ionele CM, et al. Periodontitis, metabolic and gastrointestinal tract diseases: Current perspectives on possible pathogenic connections. Journal of Persian Medicine. 2022;**12**(3):341

[25] Rinčić G, Gaćina P, Virović Jukić L, Rinčić N, Božić D, Badovinac A. Association between periodontitis and liver disease. Acta Clinica Croatica. 2022;**60**(3):510-518

[26] Minihane AM, Vinoy S, Russell WR, Baka A, Roche HM, Tuohy KM, et al. Low-grade inflammation, diet composition and health: Current research evidence and its translation. The British Journal of Nutrition. 2015;**114**(7):999-1012

[27] Papic N, Samadan L, Vrsaljko N, Radmanic L, Jelicic K, Simicic P, et al. Distinct cytokine profiles in severe COVID-19 and non-alcoholic fatty liver disease. Life (Basel). 2022;**12**(6):795

[28] Tilg H, Moschen AR. Evolution of inflammation in nonalcoholic fatty liver disease: The multiple parallel hits hypothesis. Hepatology. 2010;**52**(5):1836-1846

[29] Surlin P, Didilescu AC, Lazar L, Arsenie CC, Camen A, Popescu DM, et al. Evaluation through the optical coherence tomography analysis of the

influence of non-alcoholic fatty liver disease on the gingival inflammation in periodontal patients. Diabetes Metabolic Syndrome Obesity. 2021;**14**:2935-2942

[30] Eslam M, Sanyal AJ, George J. International consensus panel. MAFLD: A consensus-driven proposed nomenclature for metabolic associated fatty liver disease. Gastroenterology. 2020;**158**(7):1999-2014.e1

[31] Wanqiu X, Zhengwei Z, Lihong Y, Bing X, Hualei X, Xiumei W, et al. Exploration of shared gene signatures and molecular mechanisms between periodontitis and nonalcoholic fatty liver disease. Frontiers in Genetics. 2022;**13**:939751

[32] Kuroe K, Furuta M, Takeuchi K, Takeshita T, Suma S, Shinagawa T, et al. Association between periodontitis and fibrotic progression of non-alcoholic fatty liver among Japanese adults. Journal of Clinical Periodontology. 2021;**48**(3):368-377

[33] Di Ciaula A, Bonfrate L, Portincasa P. The role of microbiota in nonalcoholic fatty liver disease. European Journal of Clinical Investigation. 2022;**52**(7):e13768

[34] Duan Y, Pan X, Luo J, Xiao X, Li J, Bestman PL, et al. Association of inflammatory cytokines with non-alcoholic fatty liver disease. Frontiers in Immunology. 2022;**13**:880298

[35] Yun CH, Jhuang JR, Tsou MT. Pericardial fat, thoracic peri-aortic adipose tissue, and systemic inflammatory marker in nonalcoholic fatty liver and abdominal obesity phenotype. Scientific Reports. 2022;**12**(1):1958

[36] Gheorghe DN, Popescu DM, Salan A, Boldeanu MV, Ionele CM, Pitru A, et al. Non-surgical periodontal therapy could improve the periodontal inflammatory status in patients with periodontitis and chronic hepatitis C. Journal of Clinical Medicine. 2021;**10**(22):5275

[37] Hadizadeh F, Faghihimani E, Adibi P. Nonalcoholic fatty liver disease: Diagnostic biomarkers. World J Gastrointest Pathophysiol. 2017;**8**(2): 11-26

[38] Niederreiter L, Tilg H. Cytokines and fatty liver diseases. Liver Research. 2018;**2**(1):14-20

[39] Fricker ZP, Pedley A, Massaro JM, Vasan RS, Hoffmann U, Benjamin EJ, et al. Liver fat is associated with markers of inflammation and oxidative stress in analysis of data from the Framingham heart study. Clinical Gastroenterology and Hepatology. 2019;**17**(6):1157-1164.e4

[40] Dali-Youcef N, Vix M, Costantino F, El-Saghire H, Lhermitte B, Callari C, et al. Interleukin-32 contributes to human nonalcoholic fatty liver disease and insulin resistance. Hepatological Communication. 2019;**3**(9):1205-1220

[41] Han MS, White A, Perry RJ, Camporez JP, Hidalgo J, Shulman GI, et al. Regulation of adipose tissue inflammation by interleukin 6. Proceedings of the National Academic Science USA. 2019;**117**(6):2751-2760

[42] Fontes-Cal TCM, Mattos RT, Medeiros NI, Pinto BF, Belchior-Bezerra M, Roque-Souza B, et al. Crosstalk between plasma cytokines, inflammation, and liver damage as a new strategy to monitoring NAFLD progression. Frontiers in Immunology. 2021;**12**:708959

[43] Islam H, Chamberlain TC, Mui AL, Little JP. Elevated Interleukin-10 levels in COVID-19: Potentiation of pro-inflammatory responses or impaired

anti-inflammatory action? Frontiers in Immunology. 2021;**12**:677008

[44] Schwabe RF, Brenner DA. Mechanisms of liver injury. I. TNF-alpha-induced liver injury: Role of IKK, JNK, and ROS pathways. American Journal of Physiology. Gastrointestinal and Liver Physiology. 2006;**290**(4):G583-G589

[45] Tian J, Zhong R, Liu C, Tang Y, Gong J, Chang J, et al. Association between bilirubin and risk of non-alcoholic fatty liver disease based on a prospective cohort study. Scientific Reports. 2016;**6**:31006

[46] Chen SL, Morgan TR. The natural history of hepatitis C virus (HCV) infection. International Journal of Medical Sciences. 2006;**3**(2):47

[47] Bataller R, Brenner DA. Liver fibrosis. The Journal of Clinical Investigation. 2005;**115**(2):209-218

[48] Sarhan MA, Pham TN, Chen AY, Michalak TI. Hepatitis C virus infection of human T lymphocytes is mediated by CD5. Journal of Virology. 2012;**86**(7):3723-3735

[49] Capone F, Guerriero E, Colonna G, Maio P, Mangia A, Castello G, et al. Cytokinome profile evaluation in patients with hepatitis C virus infection. World Journal of Gastroenterology. 2014;**20**(28):9261-9269

[50] Coates EA, Brennan D, Logan RM. Hepatitis C infection and associated oral health problems. Australian Dental Journal. 2000;**45**(2):108-114

[51] Carozzo M. Oral health in patients with hepatitis C virus infection: An underestimated problem? Oral Diseases. 2001;**7**(5):267-270

[52] Alavian SM, Mahboobi N, Mahboobi N, Karayiannis P. Oral conditions associated with hepatitis C virus infection. The Saudi Journal of Gastroenterology. 2013;**19**(6):245-251

[53] Cacoub P, Comarmond C, Domont F, Savey L, Desbois AC, Saadoun D. Extrahepatic manifestations of chronic hepatitis C virus infection. Therapeutic Advances in Infectious Disease. 2016;**3**(1):3-14

[54] Yoon JC, Yang CM, Song Y, Lee JM. Natural killer cells in hepatitis C: Current progress. World Journal of Gastroenterology. 2016;**22**:1449-1460

[55] Wang J, Zhao P, Guo H, Sun X, Jiang Z, Xu L, et al. Serum IL 33 levels are associated with liver damage in patients with chronic hepatitis C. Mediators of Inflammation. 2012;**2012**:819636

[56] Sharma A, Chakraborti A, Das A, Dhiman RK, Chawla Y. Elevation of interleukin 18 in chronic hepatitis C: Implications for hepatitis C virus pathogenesis. Immunology. 2009;**128**(S1):e514-e522

[57] Tawfik AK, Amin AM, Yousef M, El Sayd NM, Elashry H, Elkadeem M, et al. IL 1α correlates with severity of hepatitis C virus related liver diseases. Journal of Inflammation Research. 2018;**11**:289-295

[58] Vanis N, Mehmedović A, Mesihović R. Use of serum levels of proinflammatory cytokine IL 1α in chronic hepatitis C. Collegium Antropologicum. 2015;**39**:75-79

[59] Surlin P, Gheorghe DN, Popescu DM, Martu AM, Solomon S, Roman A, Lazar L, Stratul S.I, Rusu D, Foia L. Interleukin-1α and -1β assessment in the gingival crevicular fluid of periodontal patients with chronic hepatitis C. Experimental and Therapeutic Medicine 2020;**20**:2381-2386

[60] Reis C, Costa DA, Guimarães JT, Tuna D, Braga AC, Pacheco JJ, et al. Clinical improvement following therapy for periodontitis: Association with a decrease in IL 1 and IL 6. Experimental and Therapeutic Medicine. 2014;**8**:323-327

[61] Jia HY, Du J, Zhu SH, Ma YJ, Chen HY, Yang BS, et al. The roles of serum IL 18, IL 10, TNF alpha and sIL 2R in patients with chronic hepatitis C. Hepatobiliary & Pancreatic Diseases International. 2002;**1**:378-382

[62] Isaza-Guzmán DM, Medina-Piedrahíta VM, Gutiérrez- Henao C, Tobón-Arroyave SI. Salivary levels of NLRP3 inflammasome-related proteins as potential biomarkers of periodontal clinical status. Journal of Periodontology. 2017;**88**(12):1329-1338

[63] García-Hernández AL, Muñoz-Saavedra AE, González-Alva P. Upregulation of proteins of the NLRP3 inflammasome in patients with periodontitis and uncontrolled type 2 diabetes. Oral Diseases. 2019;**25**(2):596-608

[64] Aral K, Berdeli E, Cooper PR. Differential expression of inflammasome regulatory transcripts in periodontal disease. Journal of Periodontology. 2020;**91**(5):606-616

[65] Isola G, Polizzi A, Santonocito S, Alibrandi A, Williams RC. Periodontitis activates the NLRP3 inflammasome in serum and saliva. Journal of Periodontology. 2022;**93**(1):135-114

[66] Chen W, Xu Y, Het L. HCV genomic RNA activates the NLRP3 inflammasome in human myeloid cells. PLoS One. 2014;**9**(1):e84953

[67] Negash AA, Ramos HJ, Crochet N. IL-1β production through the NLRP3 inflammasome by hepatic macrophages links hepatitis C virus infection with liver inflammation and disease. PLoS Pathogens. 2013;**9**:e1003330

[68] Burdette D, Haskett A, Presser L, McRae S, Iqbal J, Waris G. Hepatitis C virus activates interleukin-1β via caspase-1-inflammasome complex. The Journal of General Virology. 2012;**93**(2):235-246

[69] Surlin P, Lazar L, Sincar C, Gheorghe DN, Popescu DM, Boldeanu VM, et al. NLRP3 inflammasome expression in gingival crevicular fluid of patients with periodontitis and chronic hepatitis C. Mediators of Inflammation. 2021;**19**:6917919

[70] Pradeep AR, Suke DK, Prasad MR. Expression of key executioner of apoptosis caspase-3 in periodontal health and disease. Journal of Investigative and Clinical Dentistry. 2016;**7**(2):174-179

[71] Cheng R, Liu W, Zhang R, Feng Y, Bhowmick NA, Hu T. Porphyromonas gingivalis-derived lipopolysaccharide combines hypoxia to induce caspase-1 activation in periodontitis. Frontiers in Cellular and Infection Microbiology. 2017;**7**:474

[72] Figueredo CM, Rescala B, Teles RP. Increased interleukin-18 in gingival crevicular fluid from periodontitis patients. Molecular Oral Microbiology. 2008;**23**(2):173-176

[73] Niu Z, Zhang P, Tong Y. Association of plasma interleukin-18 levels and polymorphisms in interleukin-18 gene with outcomes of hepatitis C virus infections: A metaanalysis. Journal of Immunoassay. 2015;**36**(3):221-232

[74] Du Clos TW. Pentraxins: Structure, function, and role in inflammation.

International Scholarly Research Notes. 2013;**2013**:379040

[75] Pradeep AR, Kathariya R, Raghavendra NM, Sharma A. Levels of pentraxin-3 in gingival crevicular fluid and plasma in periodontal health and disease. Journal of Periodontology. 2011;**82**:734-741

[76] Mohan M, Jhingran R, Bains VK, Gupta V, Madan R, Rizvi I, et al. Impact of scaling and root planing on C-reactive protein levels in gingival crevicular fluid and serum in chronic periodontitis patients with or without diabetes mellitus. Journal of Periodontalogical Implantation Science. 2014;**44**:158-168

[77] Pradeep AR, Manjunath RG, Kathariya R. Progressive periodontal disease has a simultaneous incremental elevation of gingival crevicular fluid and serum CRP levels. Journal of Investigative and Clinical Dentistry. 2010;**1**:133-138

[78] Bian Y, Liu C, Fu Z. Application value of combination therapy of periodontal curettage and root planing on moderate-to-severe chronic periodontitis in patients with type 2 diabetes. Head & Face Medicine. 2021;**17**:12

[79] Sun J, Zheng Y, Bian X, Ge H, Wang J, Zhang Z. Non-surgical periodontal treatment improves rheumatoid arthritis disease activity: A meta-analysis. Clinical Oral Investigations. 2021;**25**:4975-4985

Chapter 6

Glance on the Critical Role of IL-23 Receptor Gene Variations in Inflammation-Induced Carcinogenesis

Mohammed El-Gedamy

Abstract

In this chapter, we will discuss the importance of genetic variations in the IL-23 receptor (IL-23R) gene in driving the process of inflammation-induced carcinogenesis. By applying bladder cancer (BLC) as a model, we will focus on two contradictory genetic mutations within the receptor gene. The first one is enhanced by cancer and induces inflammation-induced carcinogenesis via up-regulating IL-23/IL-17 inflammatory axis. However, the other preventive one deregulates this inflammatory pathway by distorting the protein nature of the receptor, leading to block its binding affinity. During the process of carcinogenesis, cancer genetically inclines the balance towards the protumor, via over-expressing the IL-23R on the surfaces of immune-bearing cells, particularly tumor-associated monocytes (TAMs) and thus increasing the levels of pro-angiogenic cytokines IL-23 and IL-17.

Keywords: bladder cancer model, IL-23 receptor, IL-23/IL-17 inflammatory axis, Tumor-associated monocytes, IL-23R gene variants

1. Introduction

Interleukin-(IL) 23 is a heterodimeric cytokine formed by a distinct p19 subunit and a shared p40 subunit with IL-12 (**Figure 1**) [2]. IL-23 engages with the heterodimeric IL-23 receptor, which consists of IL-23R chain and IL-12Rβ1 chains, so as to activate intracellular Janus kinases (JAKs), mainly tyrosine kinase-2 (TYK2) and JAK2, which stimulates signal transducer and activator of transcription (STAT3). Subsequently, the JAKS-STAT-3 pathway plays a critical role in the upstreaming genes of a variety of proinflammatory cytokines, such as IL-6 and IL-17 [3].

Antigen-presenting cells (APCs), such as DCs and macrophages, are thought to be the predominant provider of both IL-23 and IL-12 cytokines [4]. It is now acknowledged that the imbalance between the inflammatory cytokine IL-23 and IL-12 in tumors can re-shape the development of pro-tumor or anti-tumor immunity. Given that the role of IL-12 in boosting anti-tumor immunity is well recognized and recently evaluated [5].

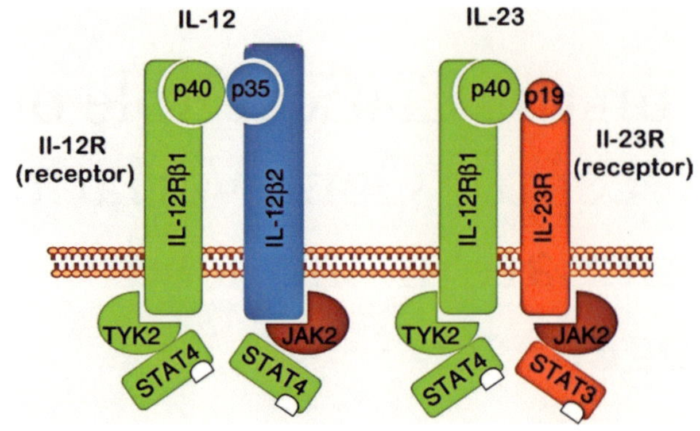

Figure 1.
Schematic representation of the interleukin (IL)-12 and IL-23 cytokine family and their receptors structures, and associated Janus kinases (JAKs) and signal transducers and activators of transcription (STATs) signaling partners [1].

Concerning common in vitro trials, mice that were deficient for IL-23p19 protein injected with chemical carcinogen-induced fibrosarcomas, showed a declined rate and frequency of tumor-growth compared to wild-type (WT) controls [6]. However, rodents lacking IL-12/23p40 or IFN-γ were found to have a significantly higher chance of developing carcinomas [7–9]. Also, using chemical carcinogen-induced skin cancer, IL-23-p19-deficient mice revealed significantly decreased numbers of cutaneous papillomas compared to WT mice, while the opposite was observed in IL-12p35-deficient mice [10].

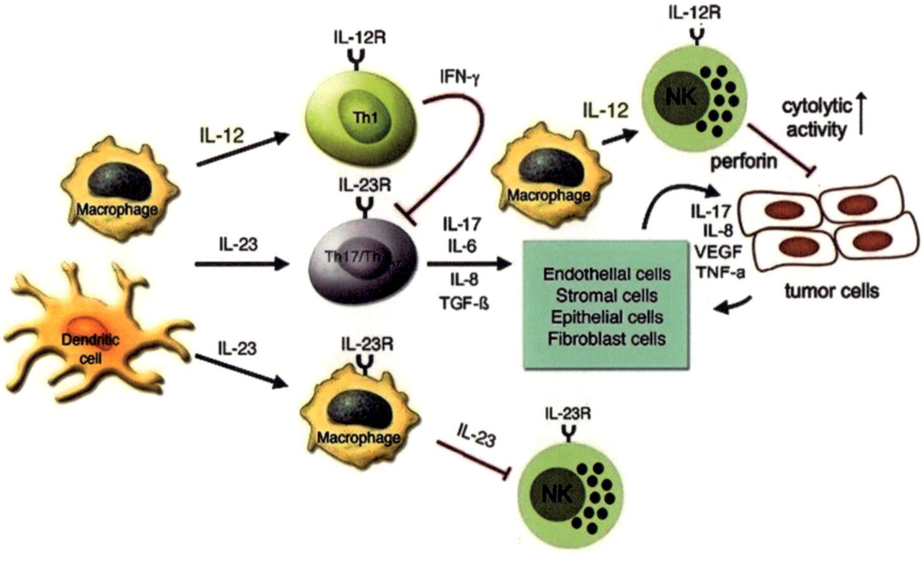

Figure 2.
Role of IL-23/IL-17 inflammatory axis in cancer. IL-23 produced by tumor-associated macrophages via autocrine and paracrine manners enhances the development of IL-17 producing T-cells called Th-17 cells and that stimulates the generation of pro-tumorigenic cytokines, such as IL-6, IL-17, VEGF, and TNF-α, thus, contributing to tumor growth. However, IL-12 produced by anti-tumor macrophages promotes, cytotoxic T-lymphocytes and natural killers to produce INF-γ and thereby killing tumor cells [11].

In vivo, IL-23 plays a vital role in inducing the effector function and proliferation of Th-17 cells (**Figure 2**), which are characterized by the expression of proinflammatory cytokine IL-17, under the influence of the master transcription factor retinoic-acid receptor-related orphan-receptor-γt (RORγt) [12–16].

Prolonged activation of the IL-23/IL-17 pathway is believed to stimulate the incidence and tumor growth [17]. This is owing to its ability to facilitate tumor growth and metastasis by up-regulating the synthesis of pro-angiogenic factors in fibroblasts and endothelial cells [18, 19].

IL-12, on the other side, mediates the anti-tumor immunity by driving the differentiation of Th-1 cells to cytotoxic T-lymphocytes and producing Interferon-gamma-(IFNγ)-II [5, 20]. In similar manner, IL-12, produced by anti-tumor macrophages, acts mainly on lymphoid cells such as NKs (**Figure 2**), to directly augment NKs proliferation and activate its cytotoxic function against tumor cells via IFN-γ secretion [5, 21, 22].

In the tumor micro-environment, it has been shown that IL-23 produced by tumor-associated monocytes (TAMs) by autocrine and paracrine manners may lead to the down-regulation of perforin, IL-12, and IFN-γ, thereby, suppressing T-cytotoxic lymphocytes and NKs effector functions and that indirectly results in enhancing tumor growth and development [6, 23].

2. IL-23 receptor

IL-23 receptor (IL-23R) is a heterodimeric construct formed by IL-23R and IL-12Rb1 subunits. IL-12Rb1 is also a part of the IL-12 receptor, however, IL-23R is restricted to the IL-23 receptor complex [24, 25].

IL-23R is mainly expressed on the surfaces of memory T-cells, NKs, macrophages, and Dendritic cells (DCs), while its expression is induced on $CD4^+$ T-cells during the differentiation towards the Th-17 cells with cellular responsiveness to IL-23 [24, 26–28].

In cancer-related inflammation, IL-23/IL-23R interaction is truly required for Th-17 cell-mediated immune response [29, 30] and represents the potential connection between the failure of the adaptive immune surveillance and tumor-promoting pro-inflammatory processes [10].

The intrinsic signaling pathway begins when IL-23 ligates to its receptor and that induces autophosphorylation and transphosphorylation of receptor-associated proteins namely Janus kinases (JAK2) and tyrosine kinases (TYK2) which are located in the intracellular domain of the receptor subunits (**Figure 3**). These phosphorylated residues serve as docking sites for the phosphorylation homodimerization of the signal transducer and activator of transcription (STAT) molecules. To detail, STAT3 is the main player in the IL-23 signaling pathway, while STAT4 is the central player in the IL-12 pathway [24, 25, 31, 32].

Once activated, the homodimers of STAT3 translocate into the nucleus, wherein they bind to the DNA in the promoter region of the target gene, such as the IL-17 gene [31, 32].

It has recently been reported that STAT3 has key role in the differentiation of T-helper-17 [33], however, STAT5 activation hinders T-helper 17-cell development [34]. For these reasons, the selective activation of certain STAT molecules may be functionally important in disease susceptibility because of the role that these molecules have in the regulation of naive $CD4^+$ T-cell differentiation [35].

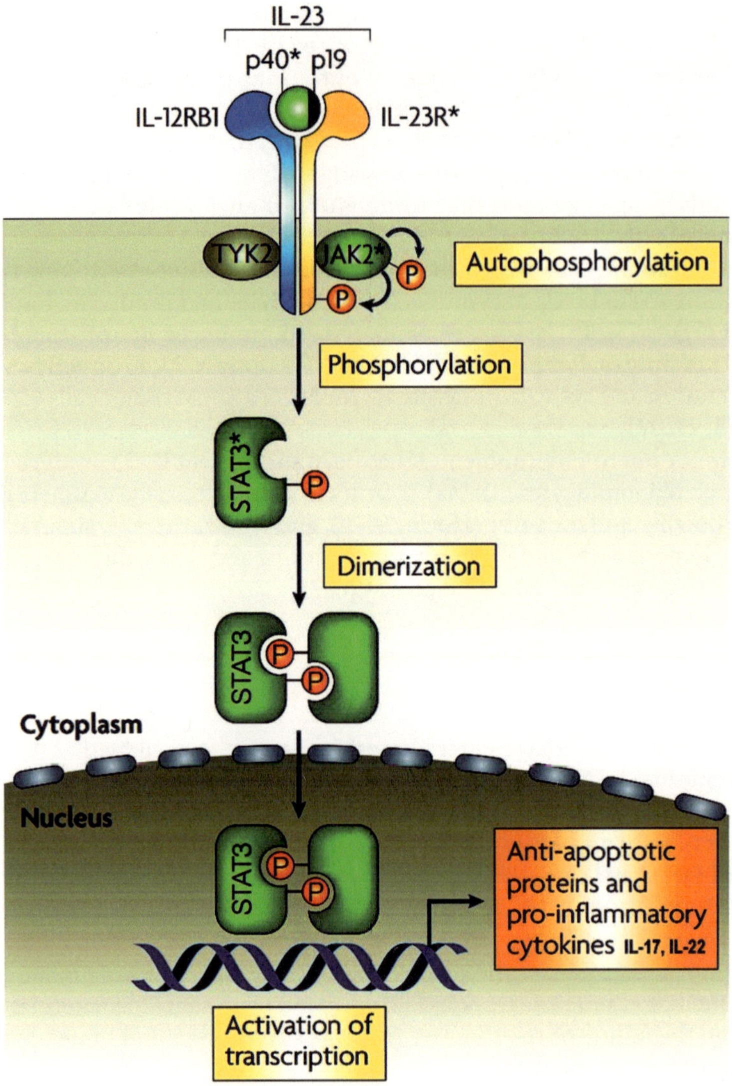

Figure 3.
IL-23 signaling pathway. Functional interleukin-23 receptor (IL-23R) signaling issued from the interaction of a heterodimeric cytokine (formed from p40 and p19 subunits) with a heterodimeric receptor (formed from IL-23R and IL-12RB1 subunits). On engagement of IL-23 with IL-23R, Janus-kinase-2 (JAK2) is activated, resulting in JAK2 autophosphorylation and tyrosine phosphorylation of the receptor. Phospho-STAT3 proteins homodimerize and translocate into the nucleus inducing transcription of anti-apoptotic proteins and cytokines, such as IL-17 and IL-22 [31, 32].

In cancer-related inflammation, malignant cells trigger systemic and local alterations in the tumor microenvironment, enabling them to evade anti-tumor immune response, and contribute to niche creation for tumor progression and metastasis. Of note, the tumor microenvironment is a multicellular system consisted of resident stromal cells (including endothelial cells, fibroblasts, and mesenchymal cells) and tumor-associated myeloid-derived cells aggregated into the extracellular matrix, which closely interact with malignant cells and contribute to tumorigenesis [36].

The tumor-infiltrating myeloid cell is composed of mast cells, TAMs, DCs, T-helper-17 (Th-17), neutrophils, natural killers (NKs), and myeloid-derived suppressors cells. These cells (excluding NKs) are driven to the primary tumor location, to sustain a permanent state of inflammation by producing pro-inflammatory mediators, such as reactive nitrogen radicals, tumor necrosis factor-α, chemokine receptor type-7, vascular endothelial growth factor (VEGF), cyclooxygenase-2 enzyme and hypoxia-inducible factor-1. These agents are shown to generate DNA damage to the tissue cells as well as drive neoplastic transformation [37–42].

IL-23, a proinflammatory cytokine, is a member of the family of heterodimeric cytokines comprising of a unique IL-23p19 subunit covalently bound to a p40 subunit shared with IL-12 [43–45].

IL-23 is synthesized mainly by macrophages and DCs as well as maintains the self-promotion of these phagocytes via IL-23 receptor (IL-23R)-mediated autocrine manner. The core role of IL-23 is to boost the differentiation of naïve $CD4^+$ T-cells and hinder T-helper (Th)-1 and Th-2 differentiation for the generation of T-helper-(Th)-17 cells [10, 46–50]. In an apparent paradox, it has been reported that IL-12 contributes to the anti-tumor immune response by mediating a polarization of the Th-1-cells to secrete interferon-γ and that increases the cytotoxic activity of both NKs and $CD8^+$ T-cells against tumor cells [51, 52].

During carcinogenesis, Th-17 cells, by expressing high amounts of IL-17, are believed to be responsible for increased the production of VEGF from fibroblast and endothelial cells that correlated with stimulating vascular endothelial proliferation, causing the induction of angiogenesis, tumor growth, and metastasis [10, 46, 47, 49].

3. IL-2R gene polymorphisms and cancer

The human IL-23R gene is situated in the short-arm of chromosome-1 [1p31.2 ~32.1] between 67,260,000–67,580,000 base-pairs and separated by 150 kb from the neighboring gene, IL12Rb2 [53]. The native form of the human IL-23R gene has 11 exons. The transcribed mRNA is translated into a protein of 629 amino acids as a constituent part of the receptor transmembrane proteins [43]. Importantly, data shows that single nucleotide polymorphisms (SNPs) in the IL-23R gene (**Figure 4**), especially rs-6682925, rs-1884444, and rs-10889677 have significant impacts on the cancer susceptibility [54].

Based on the NCBI SNP database, it was reported that the rs6682925 T/C variant is correlated with a higher risk of hepatocellular carcinoma [55], esophageal cancer [56] and acute myeloid leukemia [57]. It is attributable to the variant located at the promoter region of IL-23R gene at 907-bp upstream from the transcriptional start position) was linked to increased reporter gene activity [55].

With respect to the rs-1884444 SNP, several case-control studies have examined the association between the variant and the risk of multiple cancers, such as hepatocellular carcinoma [55, 58], esophageal [56, 59, 60], gastric cancer [61] acute myeloid leukemia [57], colorectal cancer [62], colorectal adenoma [62, 63], colon cancer [63], rectal cancer [63], lung cancer [64], nasopharyngeal cancer [64], breast cancer [64, 65], and acute lymphoblastic leukemia [66].

Rs-1884444 "G/T" is a non-synonymous single nucleotide polymorphism (SNP) positioned at the exon-2 in the coding region of the IL-23R gene, that causes the amino acid glutamine (Gln) to be replaced by histidine (His) in the signal-peptide at the extracellular domain of IL-23R, influencing its specificity and affinity to IL-23 [61, 67].

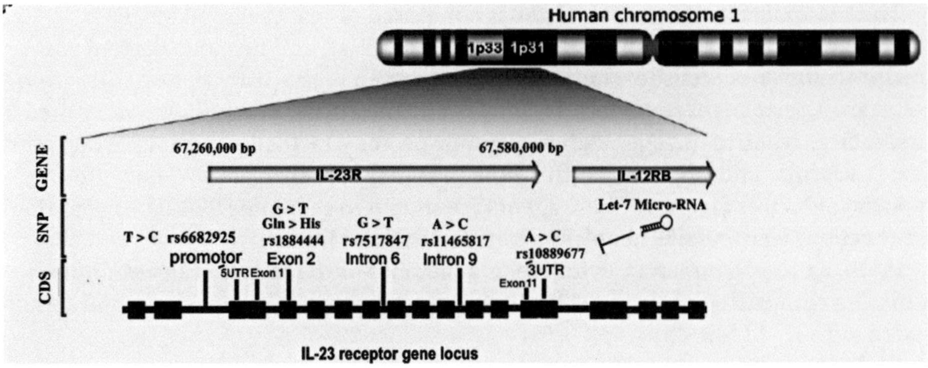

Figure 4.
Schematic diagram of the IL-23 receptor SNPs in cancer. A total of 6 SNPs were genotyped across a 320 kb region and were reported to associate with cancer incidence. Positions and gene boundaries were determined by the NCBI database. Abbreviations; CDS, Coding sequence. SNP, single nucleotide polymorphism.

Previous reports indicated that rs-10889677 "A/C" SNP is correlated with genetic susceptibility to various tumors, such as gastric cancer [68–70], oral cancer [71], colorectal cancer [72], hepatocellular carcinoma [73], bladder cancer [74], breast cancer [65], ovarian cancer [75], lung and nasopharyngeal cancers [64]. That is because the rs-10889677 A/C polymorphism resides in the 3′-untranslated region (3′-UTR) of the IL-23R, results in the substitution of nucleobase adenine (A) by cytosine (C), which could increase the binding affinity of micro-RNA let-7f and, thus, elevated the transcription of the IL-23R gene in vitro and in vivo [64]. Eventually, Zhang and co-workers indicated that higher frequencies of both rs11465817 "A" and rs7517847 "G" alleles were associated with advanced ovarian cancer [75].

4. Impact of IL-23R gene polymorphisms on reshaping tumor micro-environment

By the ligation with the IL-23 receptor (IL-23R), IL-23 exerts its biological activities on the surface of target cells [43–45].

Under cancer-causing inflammation, we embarked to discuss in this chapter the potential relationship between SNPs rs-10889677 "A/C" and rs-1884444 "G/T" in IL-23R and the susceptibility to and progression of cancer.

When proposed bladder cancer (BLC) as a model, Gedamy et al. [76] found that the silent variant rs-10889677 "A/C" has been closely correlated with the development and progression of BLC. This was obvious because people bearing the "C" allele were more susceptible to develop BLC, compared with individuals carrying "A" allele. In addition, they showed a notable increase in the expression levels of IL-23R in the sera of BLC patients as a result of the SNP rs10889677 "A/C" variation.

The possible rationale is that the rs10889677 SNP lies within a predicted binding site for human mi-let-7f [64, 65]. This micro-RNA would therefore lose its binding ability to IL-23R mRNA transcripts bearing the rs10889677C allele, resulting in the up-regulation of IL-23R expression. Amazingly, Gedamy et al. [76] observed a corresponding uptick in serum levels of IL-23 and IL-17 as a logical consequence, highlighting the importance of the rs10889677 "A/C" polymorphism in driving the prolonged activation of IL-23/17 inflammatory axis and thus beefing up the mechanism of inflammation-related carcinogenesis [18, 19, 47, 64, 65].

In the process of carcinogenesis, functional interactions could probably occur between infiltrating TAMs and cancer cells, to create a pathway for augmenting IL-23 and IL-17 production, and ultimately leads to tumor progression and metastases [49, 77, 78]. To generate TAMs during the process of carcinogenesis, IL-23 originated from TAMs requires to act on them via the IL-23R-mediated autocrine way, by which this mechanism of stimulation is largely reliant on the amounts of IL-23R on the surfaces of these cells [24, 26, 43, 79].

In response to the SNP rs-10889677 variation (enhanced by cancer), up-regulation of IL-23 receptor on TAMs surfaces and more cellular auto-activation can be acquired. In this situation, a virtuous loop emerges in which the production of the pro-tumor cytokine IL-23 is increased [24, 26, 43, 79].

Following that, two functional interactions between TAM-expressed IL-23 and other receptor-bearing cells are anticipated to occur (**Figure 5**). First, researchers have

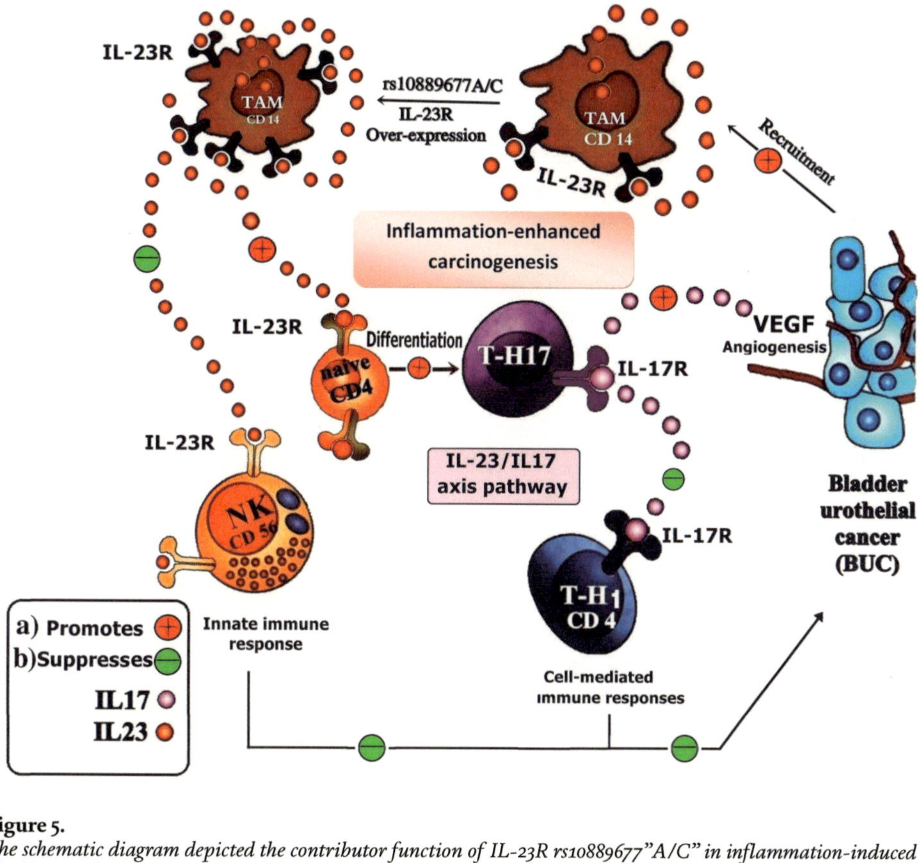

Figure 5.
The schematic diagram depicted the contributor function of IL-23R rs10889677"A/C" in inflammation-induced Bladder urothelial carcinogenesis. First, rs10889677 variant impairs transcription binding sites of microRNA (miR-let-7f) to 3'UTR of the IL-23R-gene, leading to the up-regulation of IL-23R expression on the surfaces of receptor-harboring immune cells including IL-23R-positive NKs, TAMs and $CD4^+$ cells. In the case of IL-23R-positive TAMs, over-expression of IL-23R causes these cells to self-activate (in an autocrine way) and enhance their production of IL-23. Second, binding to the IL-23R complex is required for IL-23 action. After binding to the receptor, IL-23 plays two contradictory roles on the IL-23R-harboring cells and such roles will increase gradually as a result of IL-23R over-expression; A) Promotion role: IL-23 promotes the differentiation and maturation of naïve CD4+ T-cells for the generation of T-helper 17 cells. The following increase in IL-17 thwarts T-helper (Th)-1 and stimulates VEGF synthesis from vascular-endothelial-cells and thus causes the enhancement of angiogenesis and metastasis. B) Suppression role: IL-23 consecutively contributes to inherent innate immuno-suppression by hindering the natural killers (such as CD56+ NK) [76].

observed that the increased amounts of IL-23 in tumor stroma may contribute to the dampening of the NK anti-tumor response by down-regulating IL-12 and IFN-γ, thereby letting tumor cells to escape from immune surveillance [6, 80].

Second, IL-23 is involved in the transformation of naive CD4+ into Th-17 cells, as well as the production of IL-17, a cytokine that mediates promotion of pro-tumor immune response; a body of evidence implies that IL-17 can directly block Th-1 cell differentiation, leading to a reduction in T-cell-mediated responses against cancer cells [81, 82]. In tumor milieu, increased levels of IL-17 also are positively inked to profuse production of a variety of pro-angiogenic factors from the endothelial cells and fibroblasts. Such pro-angiogenic factors, including VEGF, Transforming growth factor beta (TGF-β) and platelet-derived growth factor (PDGF) mediate the neo-vascularization process requisited by malignant cells for invasion and spread [10, 46, 47, 49, 83–88].

Briefly, the rs10889677 variant, by up-regulating IL-23 receptors on the surface of TAMs stimulating overproduction of IL-23, governs an indirect pro-tumor role either by maintaining the survival and expansion of Th17-cells for augmenting the expression of IL-17, or suppressing innate (e.g., NKs) and adaptive (e.g., T-helper cells) immune cells [48, 50]. This gradually leads to a pro-tumorigenic induced environment more prone to creating malignancy [10, 29, 30].

As the body's defense against the risk of tumor-associated chronic inflammation, it is biologically reasonable that genetic and epigenetic variations in IL-23R gene that result in altered IL-23R expression, structure, and/or function may play a role against cancer pathogenesis. In support of this perception, two studies observed a close association of a non-synonymous "G/T" SNP in the IL-23R (Gln3His; rs-1884444) with a reduced risk of BLC [89] and hepatocellular carcinoma [55, 58].

Evidences suggest that the protective effect of the rs-1884444 G/T variant might be directly exerted on BLC by dysregulation of the IL-23/IL-17 inflammatory axis, which is evidenced by as a substantial drop in the levels of IL-23 and IL-17 under all tested genetic models. This clearly shows that the non-conservative G/T genetic variation may disrupt the receptor's ligand-binding ability, hence mitigating the proinflammatory effects of the IL-23/IL-17 axis and lessening the process of inflammation-related carcinogenesis [90].

Considering to the rs-1884444 SNP, Gedamy et al. [89] unveiled the protective role of "G/T" variant, enhanced by the host's immune surveillance, against the BLC risk. Because the "G/T" variation occurs inside the coding region of the IL-23R gene, potential IL-23 binding sites on the receptor can be altered, causing the receptor to either fail to bind IL-23 or radically diminish binding, and thereby TAMs will be restricted from the production of IL-23 [61, 67]. That would explain the underlying cause for the decline in serum levels of IL-23 and IL-17 accompanied by such non-conservative polymorphism.

As we outlined in **Figure 6**, the interaction between the rs-1884444"G/T" variant and IL-23R is required for exerting a decisive anti-tumor effects on certain cell types bearing IL-23R on their surface, involving either to enhance adaptive (e.g., T-helper cells) and innate immune cells (e.g., NK cells) with anti-tumor influences, or alternatively impede the activation of suppressive immuno-regulatory cells (e.g., TAM). Therefore, It is plausible to believe that this variant exerts its protective activity against cancer risk mainly by disrupting IL-23/IL-23R binding function on the surface of TAMs, resulting in driving abortive activation of these cells.

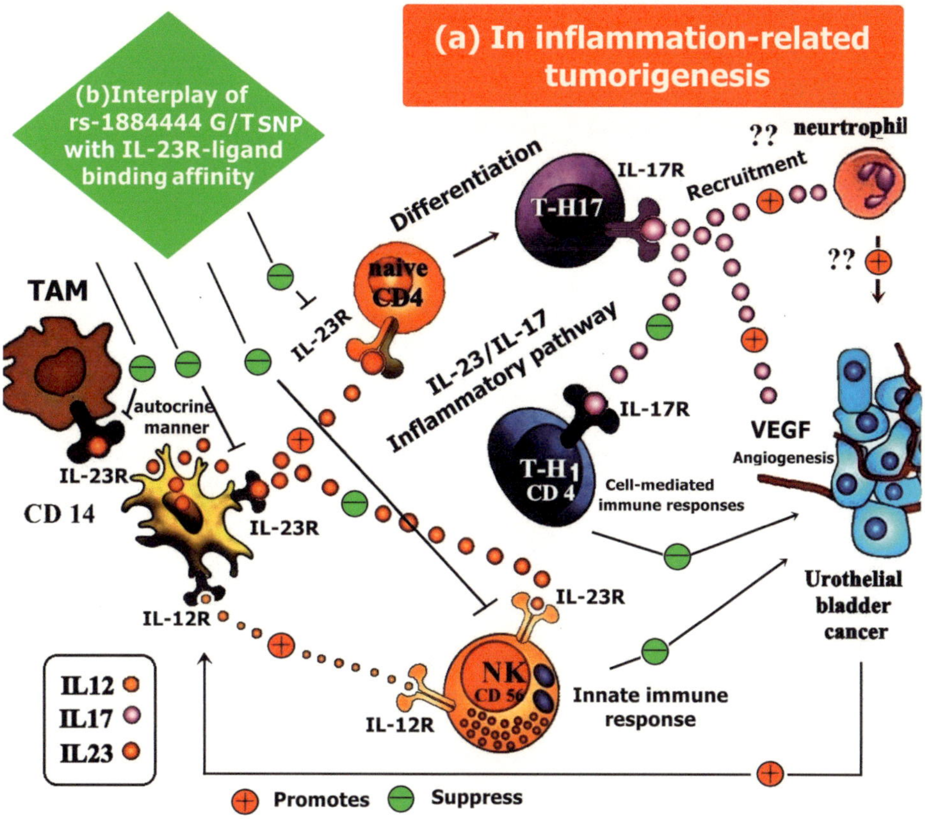

Figure 6.
Schematic diagram depicted the protective function of IL-23R rs-1884444 "G/T" variation against bladder carcinogenesis. (a) In inflammation-related tumorigenesis. Through IL-23R-mediated autocrine manner, an immune response promotes the expression of proinflammatory mediators involving IL-23 from antigen-presenting-cells seduced by cancer, such as tumor-associated monocytes (CD14⁺ TAMs) and dendritic cells (CD14+). IL-23 function is mediated by binding to the IL-23R complex. After binding to the receptor, IL-23 plays two opposing roles (suppression and promotion) on the IL-23R harboring cells; IL-23 promotes the maturation of naïve CD4+ T-cells for the production of Th-17 cells. The subsequent increase in IL-17 levels impairs T-helper (Th)-1 differentiation (cell-mediated immune response) and induces VEGF synthesis from vascular-endothelial-cells, causing the induction of tumor angiogenesis and metastasis. IL-23 contributes to innate immuno-suppression by limiting the natural killers (such as CD56+ NK) function. (b) Interplay of rs-1884444 "G/T" variant with IL-23R-ligand binding affinity. This non-conservative polymorphism in the IL-23R gene causes glutamine (Gln) to histidine (His) conversion, which modifies the IL-23R-IL-23 binding function on the surface of IL-23R-positive cells. Anti-tumor influence of the G/T variant is attained by attenuating the TAMs-IL-23/TH17-IL17 pro-tumor inflammatory axis and antagonizing the inhibitory role of IL-23 on NKs, thus enhancing the NK anti-tumor activity against tumor cells. Notably, IL-17 induces the flock of neutrophils to the neoplastic tissue, but its role in the process of tumorigenesis is still unclear [89].

The imbalance between the protective mechanism and anti-tumor mechanism plays a critical role in inclining the inflammatory tumor micro-environment towards the protumor profile. Eventually, these insights targeting the IL-23R are fostering new (immune)-therapeutic approaches for cancer treatment.

Author details

Mohammed El-Gedamy
Clinical Biochemistry and Molecular Biology, Urology and Nephrology Center, Mansoura University, Egypt

*Address all correspondence to: mselgedamy@mans.edu.eg

References

[1] Yan J, Smyth MJ, Teng MW. Interleukin (IL)-12 and IL-23 and their conflicting roles in cancer. Cold Spring Harbor Perspectives in Biology. 2018;**10**:a028530

[2] Oppmann B, Lesley R, Blom B, Timans JC, Xu Y, Hunte B, et al. Novel p19 protein engages IL-12p40 to form a cytokine, IL-23, with biological activities similar as well as distinct from IL-12. Immunity. 2000;**13**:715-725

[3] Seif F, Khoshmirsafa M, Aazami H, Mohsenzadegan M, Sedighi G, Bahar M. The role of JAK-STAT signaling pathway and its regulators in the fate of T helper cells. Cell Communication and Signaling. 2017;**15**:23

[4] Hunter CA. New IL-12-family members: IL-23 and IL-27, cytokines with divergent functions. Nature Reviews Immunology. 2005;**5**:521-531

[5] Tugues S, Burkhard S, Ohs I, Vrohlings M, Nussbaum K, Vom Berg J, et al. New insights into IL-12-mediated tumor suppression. Cell Death & Differentiation. 2015;**22**:237-246

[6] Teng MWL, Andrews DM, McLaughlin N, von Scheidt B, Ngiow SF, Möller A, et al. IL-23 suppresses innate immune response independently of IL-17A during carcinogenesis and metastasis. Proceedings of the National Academy of Sciences. 2010;**107**:8328-8333

[7] Kaplan DH, Shankaran V, Dighe AS, Stockert E, Aguet M, Old LJ, et al. Demonstration of an interferon γ-dependent tumor surveillance system in immunocompetent mice. Proceedings of the National Academy of Sciences. 1998;**95**:7556-7561

[8] Koebel CM, Vermi W, Swann JB, Zerafa N, Rodig SJ, Old LJ, et al. Adaptive immunity maintains occult cancer in an equilibrium state. Nature. 2007;**450**:903-907

[9] Smyth MJ, Thia KY, Street SE, Cretney E, Trapani JA, Taniguchi M, et al. Differential tumor surveillance by natural killer (NK) and NKT cells. The Journal of Experimental Medicine. 2000;**191**:661-668

[10] Langowski JL, Zhang X, Wu L, Mattson JD, Chen T, Smith K, et al. IL-23 promotes tumour incidence and growth. Nature. 2006;**442**:461-465

[11] Iwakura Y, Ishigame H. The IL-23/IL-17 axis in inflammation. Journal of Clinical Investigation. 2006;**116**:1218-1222

[12] Aggarwal S, Ghilardi N, Xie M-H, de Sauvage FJ, Gurney AL. Interleukin-23 promotes a distinct CD4 T cell activation state characterized by the production of interleukin-17. Journal of Biological Chemistry. 2003;**278**:1910-1914

[13] Langrish CL, Chen Y, Blumenschein WM, Mattson J, Basham B, Sedgwick JD, et al. IL-23 drives a pathogenic T cell population that induces autoimmune inflammation. Journal of Experimental Medicine. 2005;**201**:233-240

[14] Langrish CL, McKenzie BS, Wilson NJ, de Waal Malefyt R, Kastelein RA, Cua DJ. IL-12 and IL-23: Master regulators of innate and adaptive immunity. Immunological Reviews. 2004;**202**:96-105

[15] Monin L, Gaffen SL. Interleukin 17 family cytokines: Signaling mechanisms, biological activities, and therapeutic

implications. Cold Spring Harbor Perspectives in Biology. 2018;**10**:a028522

[16] Patel DD, Kuchroo VK. Th17 cell pathway in human immunity: Lessons from genetics and therapeutic interventions. Immunity. 2015;**43**:1040-1051

[17] Kolls JK, Lindén A. Interleukin-17 family members and inflammation. Immunity. 2004;**21**:467-476

[18] Numasaki M, Fukushi J-i, Ono M, Narula SK, Zavodny PJ, Kudo T, et al. Interleukin-17 promotes angiogenesis and tumor growth. Blood. 2003;**101**:2620-2627

[19] Takahashi H, Numasaki M, Lotze MT, Sasaki H. Interleukin-17 enhances bFGF-, HGF-and VEGF-induced growth of vascular endothelial cells. Immunology Letters. 2005;**98**:189-193

[20] Colombo MP, Trinchieri G. Interleukin-12 in anti-tumor immunity and immunotherapy. Cytokine & Growth Factor Reviews. 2002;**13**:155-168

[21] Cao X, Leonard K, Collins LI, Cai SF, Mayer JC, Payton JE, et al. Interleukin 12 stimulates ifn-γ–mediated inhibition of tumor-induced regulatory T-cell proliferation and enhances tumor clearance. Cancer Research. 2009;**69**:8700-8709

[22] Chan SH, Perussia B, Gupta JW, Kobayashi M, Pospisil M, Young HA, et al. Induction of interferon gamma production by natural killer cell stimulatory factor: Characterization of the responder cells and synergy with other inducers. The Journal of Experimental Medicine. 1991;**173**:869-879

[23] Langowski JL, Kastelein RA, Oft M. Swords into plowshares: IL-23 repurposes tumor immune surveillance. Trends in Immunology. 2007;**28**:207-212

[24] Parham C, Chirica M, Timans J, Vaisberg E, Travis M, Cheung J, et al. A receptor for the heterodimeric cytokine IL-23 is composed of IL-12Rbeta1 and a novel cytokine receptor subunit, IL-23R. Journal of Immunology. 2002;**168**:5699-5708

[25] Zou J, Presky DH, Wu C-Y, Gubler U. Differential associations between the cytoplasmic regions of the interleukin-12 receptor subunits β1 and β2 and JAK kinases. Journal of Biological Chemistry. 1997;**272**:6073-6077

[26] Belladonna ML, Renauld J-C, Bianchi R, Vacca C, Fallarino F, Orabona C, et al. IL-23 and IL-12 have overlapping, but distinct, effects on murine dendritic cells. The Journal of Immunology. 2002;**168**:5448-5454

[27] Caspi RR, Rachitskaya AV, Horai R, Li Z, Luger D, Villasmil R, et al. NKT cells constitutively express IL-23 receptor and RORγt, and rapidly produce IL-17 upon receptor ligation in an IL-6-independent fashion. FASEB Journal. 2008;**22**:1069.5

[28] Teng MW, Bowman EP, McElwee JJ, Smyth MJ, Casanova JL, Cooper AM, et al. IL-12 and IL-23 cytokines: From discovery to targeted therapies for immune-mediated inflammatory diseases. Nature Medicine. 2015;**21**:719-729

[29] Chen Z, Laurence A, O'Shea JJ. Signal transduction pathways and transcriptional regulation in the control of Th17 differentiation. Seminars in Immunology Journal. Academic Press. 2007;**19**:400-408

[30] Volpe E, Servant N, Zollinger R, Bogiatzi SI, Hupé P, Barillot E, et al. A critical function for transforming

growth factor-β, interleukin 23 and proinflammatory cytokines in driving and modulating human T H-17 responses. Nature Immunology. 2008;**9**:650

[31] Cho JH. The genetics and immunopathogenesis of inflammatory bowel disease. Nature Reviews Immunology. 2008;**8**:458-466

[32] Di Cesare A, Di Meglio P, Nestle FO. The IL-23/Th17 Axis in the Immunopathogenesis of Psoriasis. Journal of Investigative Dermatology. 2009;**129**:1339-1350

[33] Watford WT, Hissong BD, Bream JH, Kanno Y, Muul L, O'Shea JJ. Signaling by IL-12 and IL-23 and the immunoregulatory roles of STAT4. Immunological Reviews. 2004;**202**:139-156

[34] Laurence A, Tato CM, Davidson TS, Kanno Y, Chen Z, Yao Z, et al. Interleukin-2 signaling via STAT5 constrains T helper 17 cell generation. Immunity. 2007;**26**:371-381

[35] Zhou L, Ivanov II, Spolski R, Min R, Shenderov K, Egawa T, et al. IL-6 programs TH-17 cell differentiation by promoting sequential engagement of the IL-21 and IL-23 pathways. Nature Immunology. 2007;**8**:967-974

[36] Horimoto Y, Polanska UM, Takahashi Y, Orimo A. Emerging roles of the tumor-associated stroma in promoting tumor metastasis. Cell Adhesion and Migration. 2012;**6**:193-202

[37] Balkwill F, Mantovani A. Inflammation and cancer: Back to Virchow? The Lancet. 2001;**357**(9255): 539-545

[38] Lin C, Lin W, Yeh S, Li L, Chang C. Infiltrating neutrophils increase bladder cancer cell invasion via modulation of androgen receptor (AR)/MMP13 signals. Oncotarget. 2015;**6**:43081-43089

[39] Nesi G, Nobili S, Cai T, Caini S, Santi R. Chronic inflammation in urothelial bladder cancer. Virchows Archiv. 2015;**467**:623-633

[40] Sui X, Lei L, Chen L, Xie T, Li X. Inflammatory microenvironment in the initiation and progression of bladder cancer. Oncotarget Oncotarget. 2017;**8**:93279-93294

[41] Williams CB, Yeh ES, Soloff AC. Tumor-associated macrophages: Unwitting accomplices in breast cancer malignancy. Npj Breast Cancer. 2016;**2**:15025

[42] Zhu Z, Shen Z, Xu C. Inflammatory pathways as promising targets to increase chemotherapy response in bladder cancer. Mediators Inflamm. 2012;**2012**:528690-528690

[43] Parham C, Chirica M, Timans J, Vaisberg E, Travis M, Cheung J, et al. A receptor for the heterodimeric cytokine IL-23 is composed of IL-12Rβ1 and a novel cytokine receptor subunit, IL-23R. The Journal of Immunology. 2002;**168**:5699

[44] Sun L, He C, Nair L, Yeung J, Egwuagu CE. Interleukin 12 (IL-12) family cytokines: Role in immune pathogenesis and treatment of CNS autoimmune disease. Cytokine. 2015;**75**:249-255

[45] Vignali DA, Kuchroo VK. IL-12 family cytokines: Immunological playmakers. Nature Immunology. 2012;**13**:722-728

[46] Crew J, O'Brien T, Bradburn M, Fuggle S, Bicknell R, Cranston D, et al. Vascular endothelial growth factor is a predictor of relapse and stage progression in superficial bladder. Cancer Research Journal. 1997;**57**:5281-5285

[47] Martin-Orozco N, Dong C. The IL-17/IL-23 axis of inflammation in cancer: Friend or foe? Current Opinion in Investigational Drugs. 2009;**10**:543-549

[48] Moschen AR, Tilg H, Raine T. IL-12, IL-23 and IL-17 in IBD: Immunobiology and therapeutic targeting. Nature Reviews. Gastroenterology & hepatology. 2019;**16**:185-196

[49] Murugaiyan G, Saha B. Protumor vs antitumor functions of IL-17. The Journal of Immunology. 2009;**183**:4169

[50] Veldhoen M, Hocking RJ, Atkins CJ, Locksley RM, Stockinger B. TGFβ in the context of an inflammatory cytokine milieu supports de novo differentiation of IL-17-producing T cells. Immunity. 2006;**24**:179-189

[51] Trinchieri G. Interleukin-12 and the regulation of innate resistance and adaptive immunity. Nature Reviews Immunology. 2003;**3**:133

[52] Watford WT, Moriguchi M, Morinobu A, Oshea JJ. The biology of IL-12: Coordinating innate and adaptive immune responses. CGFR. Cytokine and Growth Factor Reviews. 2003;**14**:361-368

[53] Duerr RH, Taylor KD, Brant SR, Rioux JD, Silverberg MS, Daly MJ, et al. A genome-wide association study identifies IL23R as an inflammatory bowel disease gene. Science Journal. 2006;**314**(5804):1461-1463

[54] Yao J, Liu L, Yang M. Interleukin-23 receptor genetic variants contribute to susceptibility of multiple cancers. Gene. 2014;**533**:21-25

[55] Xu Y, Liu Y, Pan S, Liu L, Liu J, Zhai X, et al. IL-23R polymorphisms, HBV infection, and risk of hepatocellular carcinoma in a high-risk Chinese population. Journal of Gastroenterology. 2013;**48**:125-131

[56] Chu H, Cao W, Chen W, Pan S, Xiao Y, Liu Y, et al. Potentially functional polymorphisms in IL-23 receptor and risk of esophageal cancer in a Chinese population. International Journal of Cancer. 2012;**130**:1093-1097

[57] Qian X, Cao S, Yang G, Pan Y, Yin C, Chen X, et al. Potentially functional polymorphism in IL-23 receptor and risk of acute myeloid leukemia in a Chinese population. PloS One. 2013;**8**:e55473

[58] Peng Q, Qin Y, Chen Z, Deng Y, Xu J, Li S, et al. Correlation between interleukin23 receptor gene polymorphisms and risk of hepatitis B virus infection in patients. Molecular Medicine Reports. 2013;**8**:613-620

[59] Li M, Yue C, Jin G, Guo H, Ma H, Wang G, et al. Rs1884444 variant in IL23R gene is associated with a decreased risk in esophageal cancer in Chinese population. Molecular Carcinogenesis. 2019;**58**:1822-1831

[60] Ni B, Chen S, Xie H, Ma H. Functional polymorphisms in interleukin-23 receptor and susceptibility to esophageal squamous cell carcinoma in Chinese population. PLoS One. 2014;**9**:e89111-e89111

[61] Chen J, Lu Y, Zhang H, Ding Y, Ren C, Hua Z, et al. A nonsynonymous polymorphism in IL23R gene is associated with risk of gastric cancer in a Chinese population. Molecular Carcinogenesis. 2010;**49**:862-868

[62] Li B, Gan A, Zhang X, Huang W, Yu Z, Chen X. Application of high resolution melting assay to explore the correlation between the single nucleotide polymorphisms of IL-23/ IL-17 gene and colorectal cancer.

New England Journal of Medicine. 2016;**47**:661-665

[63] Poole EM, Curtin K, Hsu L, Duggan DJ, Makar KW, Xiao L, et al. Genetic variability in IL23R and risk of colorectal adenoma and colorectal cancer. Cancer Epidemiology. 2012;**36**:e104-e110

[64] Zheng J, Jiang L, Zhang L, Yang L, Deng J, You Y, et al. Functional genetic variations in the IL-23 receptor gene are associated with risk of breast, lung and nasopharyngeal cancer in Chinese populations. Carcinogenesis. 2012;**33**:2409-2416

[65] Wang L, Liu W, Jiang W, Lin J, Jiang Y, Li B, et al. A miRNA binding site single-nucleotide polymorphism in the 3'-UTR region of the IL23R gene is associated with breast cancer. PLoS One. 2012;**7**:e49823

[66] Zareinejad M, Samiei A, Valibeigi B, Gholami T, Zareifar S, Amirghofran Z. Interleukin-23 receptor gene variants in acute lymphoblastic leukemia and their relation to prognostic factors. Iranian Journal of Immunology. 2017;**14**:59-72

[67] Kan SH, Mancini G, Gallagher G. Identification and characterization of multiple splice forms of the human interleukin-23 receptor alpha chain in mitogen-activated leukocytes. Genes and Immunity. 2008;**9**:631-639

[68] Chen B, Zeng Z, Xu L, Wu X, Yu J, Xue L, et al. IL23R +2199A/C polymorphism is associated with decreased risk of certain subtypes of gastric cancer in Chinese: A case-control study. Cancer Epidemiology. 2011;**35**:165-169

[69] Dong K, Xu Y, Yang Q, Shi J, Jiang J, Chen Y, et al. Associations of functional MicroRNA binding site polymorphisms in IL23/Th17 inflammatory pathway genes with gastric cancer risk. Mediators of Inflammation. 2017;**2017**:6974696

[70] Jia Z-F, Cao D-H, Wu Y-H, Jin M-S, Pan Y-C, Cao X-Y, et al. Lethal-7-related polymorphisms are associated with susceptibility to and prognosis of gastric cancer. World Journal of Gastroenterology. 2019;**25**:1012

[71] Chien MH, Hsin CH, Chou LS, Chung TT, Lin CH, Weng MS, et al. Interleukin-23 receptor polymorphism as a risk factor for oral cancer susceptibility. Head Neck. 2012;**34**:551-556

[72] Mosallaei M, Simonian M, Esmaeilzadeh E, Bagheri H, Miraghajani M, Salehi AR, et al. Single nucleotide polymorphism rs10889677 in miRNAs Let-7e and Let-7f binding site of IL23R gene is a strong colorectal cancer determinant: Report and meta-analysis. Cancer Genetics. 2019;**239**:46-53

[73] Peng Q, Qin Y, Chen Z, Deng Y, Xu J, Li S, et al. Correlation between interleukin-23 receptor gene polymorphisms and risk of hepatitis B virus infection in patients. Molecular Medicine Reports. 2013;**8**:613-620

[74] Tang T, Xue H, Cui S, Gong Z, Fei Z, Cheng S, et al. Association of interleukin-23 receptor gene polymorphisms with risk of bladder cancer in Chinese. Fam Cancer. 2014;**13**:619-623

[75] Zhang Z, Zhou B, Zhang J, Chen Y, Lai T, Yan L, et al. Association of interleukin-23 receptor gene polymorphisms with risk of ovarian cancer. Cancer Genetics and Cytogenetics. 2010;**196**:146-152

[76] El-Gedamy M, El-khayat Z, Abol-Enein H, El-said A, El-Nahrery E. Rs-10889677 variant in interleukin-23 receptor may contribute to creating an inflammatory milieu more susceptible to bladder tumourigenesis: Report and meta-analysis. Immunogenetics Journal. 2021;**73**:207-226

[77] Kortylewski M, Xin H, Kujawski M, Lee H, Liu Y, Harris T, et al. Regulation of the IL-23 and IL-12 balance by Stat3 signaling in the tumor microenvironment. Cancer Cell Journal. 2009;**15**(2):114-123

[78] Schön MP, Erpenbeck L. The interleukin-23/interleukin-17 axis links adaptive and innate immunity in psoriasis. Frontiers in Immunology Journal. 2018;**9**:1323

[79] Grohmann U, Bianchi R, Belladonna ML, Vacca C, Silla S, Ayroldi E, et al. IL-12 acts selectively on CD8α- dendritic cells to enhance presentation of a tumor peptide in vivo. The Journal of Immunology. 1999;**163**(6):3100-3105

[80] Konjević GM, Vuletić AM, Martinović KMM, Larsen AK, Jurišić VB. The role of cytokines in the regulation of NK cells in the tumor environment. Cytokine Journal. 2019;**117**:30-40

[81] Nakae S, Iwakura Y, Suto H, Galli SJ. Phenotypic differences between Th1 and Th17 cells and negative regulation of Th1 cell differentiation by IL-17. Journal of Leukocyte Biology. 2007;**81**(5):1258-1268

[82] O'Connor W Jr, Kamanaka M, Booth CJ, Town T, Nakae S, Iwakura Y, et al. A protective function for interleukin 17A in T cell–mediated intestinal inflammation. Nature Immunology Journal. 2009;**10**(6):603-609

[83] Kryczek I, Wei S, Zou L, Altuwaijri S, Szeliga W, Kolls J, et al. Cutting edge: Th17 and regulatory T cell dynamics and the regulation by IL-2 in the tumor microenvironment. The Journal of Immunology. 2007;**178**(11):6730-6733

[84] Chen X, Wan J, Liu J, Xie W, Diao X, Xu J, et al. Increased IL-17-producing cells correlate with poor survival and lymphangiogenesis in NSCLC patients. Lung Cancer Journal. 2010;**69**(3):348-354

[85] Chi LJ, Lu HT, Li GL, Wang XM, Su Y, Xu WH, et al. Involvement of T helper type 17 and regulatory T cell activity in tumour immunology of bladder carcinoma. Clinical & Experimental Immunology Journal. 2010;**161**(3):480-489

[86] Iida T, Iwahashi M, Katsuda M, Ishida K, Nakamori M, Nakamura M, et al. Tumor-infiltrating CD4+ Th17 cells produce IL-17 in tumor microenvironment and promote tumor progression in human gastric cancer. Oncology Reports Journal. 2011;**25**(5):1271-1277

[87] Yang B, Kang H, Fung A, Zhao H, Wang T, Ma D. The role of interleukin 17 in tumour proliferation, angiogenesis, and metastasis. Mediators of Inflammation Journal. 2014:623759

[88] Miyashita M, Sasano H, Tamaki K, Hirakawa H, Takahashi Y, Nakagawa, et al. Prognostic significance of tumor-infiltrating CD8+ and FOXP3+ lymphocytes in residual tumors and alterations in these parameters after neoadjuvant chemotherapy in triple-negative breast cancer: A retrospective multicenter study. Breast Cancer Research Journal. 2015;**17**(1):1-13

[89] El-Gedamy M, El-Khayat Z, Abol-Enein H, El-Said A, El-Nahrery E. Rs-1884444 G/T variant in IL-23 receptor is likely to modify risk of bladder urothelial carcinoma by regulating IL-23/IL-17 inflammatory pathway. Cytokine. 2021;**138**:155355

[90] Akbay EA, Koyama S, Liu Y, Dries R, Bufe LE, Silkes M, et al. Interleukin-17A promotes lung tumor progression through neutrophil attraction to tumor sites and mediating resistance to PD-1 blockade. Journal of Thoracic Oncology. 2017;**12**:1268-1279

Chapter 7

Interleukin-5 and Interleukin-5 Receptor Polymorphism in Asthma

Raghdah Maytham Hameed, Haidar Abdul Amir Najim Abood and Mohanad Mohsin Ahmed

Abstract

Asthma is a common chronic inflammatory disease of the airways of the lungs, in the world. It's associated with type 2 cytokines interleukin-4, IL-5, and IL-13, which promote airway eosinophilia, bronchial hyperresponsiveness, mucus overproduction, and immunogloubulin E synthesis. IL-5 is a cytokine known to play major role in the regulation of eosinophil formation, maturation, survival, and recruitment. Hence, an increased production of IL-5 may be contributed to the pathogenesis of asthma. The expression of human IL-5 receptor presented on eosinophils, basophils, and mast cells. Hence, a polymorphism in IL-5 receptor may be implicated in the development of asthma. Many candidate genes that could potentially contribute to the susceptibility to the disease have not been investigated to date, and not all of the polymorphisms of the candidate genes have been tested for a possible association with the disease. Taking this into consideration, IL-5 (together with the IL-5 receptor) polymorphism deserves attention as the subject of further investigations into asthma. In this review, we will address the role of IL-5 and IL-5 receptor polymorphism in asthma, describe the impact of these polymorphisms on the Blood parameters and clinical parameters. Further, give an overview of preclinical and clinical studies targeting the IL-5 and IL-5 receptor pathway.

Keywords: bronchial asthma, eosinophil, Interleukin 5, Interleukin 5 receptor, Total IgE

1. Introduction

Asthma is the most common chronic lower respiratory tract and non-communicable diseases in children and adults throughout the world, characterized by inflammation which affects both proximal and distal airways [1–5]. Asthma is involving an abnormality of airway function, specifically to wide variations in airflow limitation over short periods of time and associated with airway obstruction, variable airflow limitation, bronchial hyper responsiveness and tissue remodeling of the airway structure [6–8].

Asthma is a separate disease entity, fails to identify a primary defining characteristic which separates it from other diseases and is long-winded [6]. The disease characterized by episodic or persistent symptoms of wheezing, dyspnea, and cough [9]. It is highly prevalent chronic inflammatory diseases of the airways, with differences in etiology, immunologic mechanisms, clinical presentation, pathogenesis, comorbidities, prognosis, and response to treatment, arising from not fully understood heterogenic gene-environment interactions, while environmental factors are important in the development of asthma, genetic factors could have a critical role in the expression of the disease, but the genetic background of bronchial asthma is complex, and it is likely that multiple genes contribute to its development both directly and through gene-gene interactions [8, 10–12].

2. Pathophysiology of asthma

2.1 Immune response

Asthma is primarily an inflammatory disorder of the airways associated with T helper type 2 (Th2), cell-dependent promotion of IgE production and recruitment of mast cells and eosinophils [13]. Allergic asthma may involve adaptive and innate, antigen-independent immune responses [14]. Th2 cell-driven inflammation is likely to represent an abnormal response to harmless airborne particles by polarized immune responses to its [15, 16].

However, some respiratory viral infections cause bronchiolitis of infancy and childhood wheeze and can exacerbate established asthma. Fundamental to innate immune responses to microbes are the interactions between pathogen-associated molecular patterns and pattern recognition receptors, which are associated with the production of type I interferon, pro-inflammatory cytokines, and the Th2 cell pathway in predisposed people [17].

Allergen exposure results in the activation of numerous cells of the immune system, include dendritic cells (DCs) and Th2 lymphocytes [18]. DCs in the airway epithelium and sub mucosa detect inhaled allergens [13]. DCs then migrate to the secondary lymphatic systems where they process and present antigens via major histocompatibility complex class II (MHC II) to T- and B-lymphocytes, leading to the proliferation of Th1 or Th2 cell types and B-lymphocytes produce IgE, which binds to high affinity FcεRI on basophils and mast cells [19]. In response to allergen presentation by airway DCs, T-helper lymphocytes of the adaptive immune system control many aspects of the disease through secretion of IL-4, IL-5, IL-13, IL-17, and IL-22, and these are counterbalanced by cytokines produced by T-regulatory cells (**Figure 1**) [20].

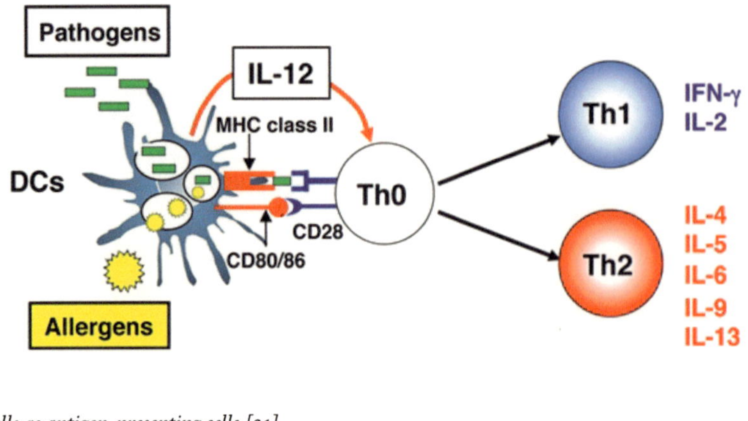

Figure 1.
Dendritic cells as antigen-presenting cells [21].

2.2 Role of T helper (Th2) cytokines in asthma

Th2-type cytokines, such as interleukin-4 (IL-4), IL-5 and IL-13, are thought to drive the disease pathology in patients with asthma and play a role in driving many of the hallmarks of allergic inflammation [22, 23]. Whereas IL-4 is important for allergic sensitization and IgE production, and IL-5 is crucial for eosinophil survival, IL-13 has pleiotropic effects in the lungs, including a central role in the development of airway hyper responsiveness (AHR) and tissue remodeling (**Figure 2**) [24].

IL-5 is the critical molecular switch for development, migration, recruiting eosinophils to the lung during allergic inflammation [26, 27]. IL-5 exerts its biological actions via stimulation of the IL-5 receptor expressed by eosinophils and, to a lesser extent, also by basophils [28].

2.3 Interleukin-5

Interleukin-5 (IL-5) is an interdigitation homodimeric glycoprotein [29]. IL-5 produced by both hematopoietic and non-hematopoietic cells including T cells, granulocytes and natural helper cells [30]. It is a T cell-derived cytokine involved in the pathogenesis of atopic diseases and play important roles in the pathogenesis of asthma, hypereosinophilic syndromes and eosinophil-dependent inflammatory diseases, through recognized as a critical regulator of eosinophilia and has effects on eosinophil progenitors, eosinophil precursors and mature eosinophils [29, 31, 32], which is believed to regulate the growth, differentiation and activation of eosinophils [33]. Further, to preferentially acts on mature eosinophils to prolong maturation and survival and increased circulating eosinophil progenitors, suggesting a key role for systemic IL-5 in eosinophil mobilization. Moreover, IL-5 causes terminal maturation of the eosinophil by increasing CCR3 expression, potentially affecting CCR3-dependent chemotaxis by eosinophils and lymphocytes [34, 35]. In addition, over

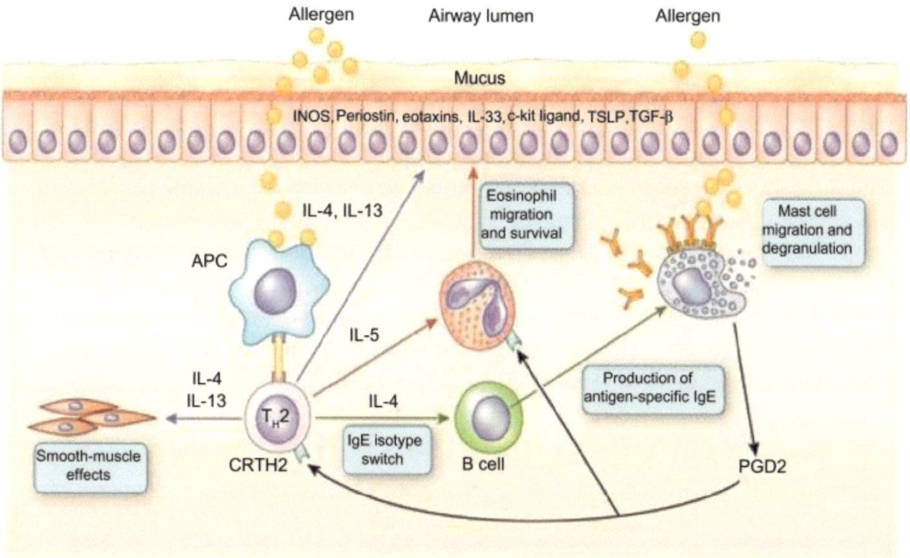

Figure 2.
Immunopathogenesis of asthma [25].

expression of IL-5 significantly increases antibody levels in vivo and reported to act as a B-cell differentiation factor by stimulating activated B cells to secrete antibody [32, 36].

Interleukin-5 is a very selective cytokine as a result of the restricted expression of the interleukin-5 receptor on eosinophils and basophils [28]. Hence, humanized monoclonal antibody to interleukin 5 significantly limited eosinophil migration to the lung [37]. Therefore; IL-5 inhibition may be an effective approach for the treatment of asthma, especially severe asthma. Interfering with eosinophil function or reducing their numbers has been one of the most important goals of therapeutic monoclonal antibodies, which target cytokine receptor interactions in asthma, particularly IL-5 [38].

IL-5 gene is located on chromosome 5 [39]. It is a potential candidate gene in the pathogenesis of asthma, may play a role in blood eosinophilia associated with atopic dermatitis [40, 41]. Locus over-expression of IL-5 significantly increases eosinophil numbers and antibody levels in vivo. Conversely, mice lacking a functional gene for IL-5 or the IL-5 receptor alpha chain (IL-5Rα) display a number of developmental and functional impairments in B-cell and eosinophil lineages [30, 32]. In addition, polymorphisms in the IL-5 genes may contribute to the susceptibility to atopic bronchial asthma and could determine the clinical course of the disease [42].

2.4 Interleukin-5 C-703T polymorphism

Interleukin-5 gene is located on chromosome 5 within a cytokine gene cluster IL-4 and IL-13 [43, 44]. Further, IL-5 gene, IL4, and IL13 may be regulated co-ordinately by long-range regulatory elements spread over 120 kilobases on chromosome 5q31.1. This gene encodes a cytokine that acts as a growth and differentiation factor for both B cells and eosinophils [44].

The IL-5 C-703 T polymorphism may play a role in blood eosinophilia associated with atopic dermatitis and it is a potential candidate gene in the pathogenesis of asthma [40, 41]. Asthma in children is associated with IL-5 C-703 T polymorphism. IL-5 C-703 T polymorphism has impact on IL-5 levels and eosinophil count. TT genotype of IL-5 C703T consider a risk factor for mild asthma in Iraqi asthmatic children [45]. In addition, IL-5 promoter polymorphism at –746A > G enhances serum total and specific IgE level responses to staphylococcal enterotoxins A, which may augment airway hyper responsiveness in adult asthmatics [46]. On the other hand, the previous results of meta-analysis suggest that IL-5 C-746 T polymorphisms are not associated with increased asthma risk, whereas IL-5 C-703 T polymorphisms are associated with asthma in children [47].

The influence and importance of IL-5 polymorphism that is single nucleotide polymorphism located within promoter regulatory sequences of cytokine gene; the promoter mutation known to cause functionally important consequences for gene expression. Although promoter mutation analysis is complex, difficult to perform, and often laborious, it is an essential part of the diagnosis of disease-causing promoter mutations and improves our understanding of the role of transcriptional regulation in human disease (**Figure 3**) [42, 45, 48].

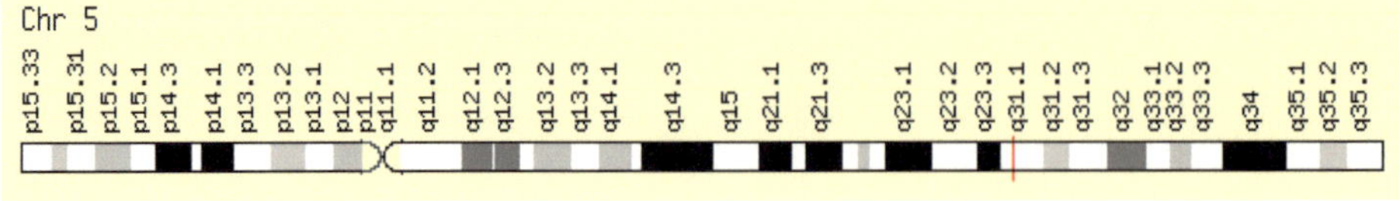

Figure 3.
Molecular location of IL-5 gene. Cytogenetic location, Chr.5q31.1 [44].

2.5 Interleukin 5 receptor (IL-5 R)

The human IL-5 receptor is a heterodimer consisting of a unique alpha subunit and a signal transducing beta subunit shared by the receptors for interleukin 3 (IL3), colony stimulating factor 2 (CSF2/GM-CSF), and interleukin 5 (IL-5) [39, 49]. The expression of human IL-5R alpha (IL-5Rα) subunit presented on eosinophils, mast cells and basophils whereas the beta subunit is more widely expressed [50].

The IL-5Rα is required for ligand-specific binding while association with the beta-chain results in increased binding affinity [51]. Further, Interleukin-5 receptor α-subunit expression may be implicated in the development of allergic diseases [52]. In addition, IL-5Rα has regulatory pathway in human eosinophils and their gene has a role for controlling eosinophils in the peripheral blood [53, 54]. It has been suggested that neutralizing antibodies to IL-5Rα could serve as a therapeutic agent in eosinophil-associated diseases [55]. An anti-interleukin 5 receptor α monoclonal antibody that depletes blood and airway eosinophils [56].

2.6 Interleukin-5 receptor alpha G-80A polymorphism

The gene for the IL-5R alpha subunit is a protein coding gene located on chromosome 3 in the region 3p24-3p26. The alternative splicing of the IL-5 Rα gene which contains 14 exons can yield several alpha IL-5 receptor isoforms [57, 58].

Polymorphisms in IL-5 Rα might be among the genetic risk factors for asthma development, especially in atopic populations [59]. Functional polymorphism in IL-5Rα may contribute to eosinophil and mast cell activation [46]. Additionally, increased expression of IL-5Rα on CD34+ cells favor eosinophilopoiesis and may thus contribute to the subsequent development of blood and tissue eosinophilia, a hallmark of allergic inflammation [60].

The G-80A polymorphisms in IL-5Rα located in the promoter region (regulatory gene region) [42]. Analyses of the activity of the IL-5Rα promoter constructs in various other eosinophils, myeloid, and non-myeloid cell lines indicated that the promoter was relatively myeloid and eosinophil lineage-specific in its expression [61]. While, previous study detected that, asthma in children is not associated with IL-5 Rα G-80A polymorphism. IL-5 R α G-80A polymorphism did not have any risk factor to develop asthma or increased asthma severity. IL-5 Rα G-80A polymorphism has not any impact on IL-5 levels, eosinophil count, and total serum IgE [62]. Other international previous studies mentioned interchangeable results explain the correlation between asthma and IL-5 Rα G-80A because of the existence of racial and ethnic specificities in the frequency distributions of the IL 5 receptor encoding polymorphic alleles (**Figure 4**) [42].

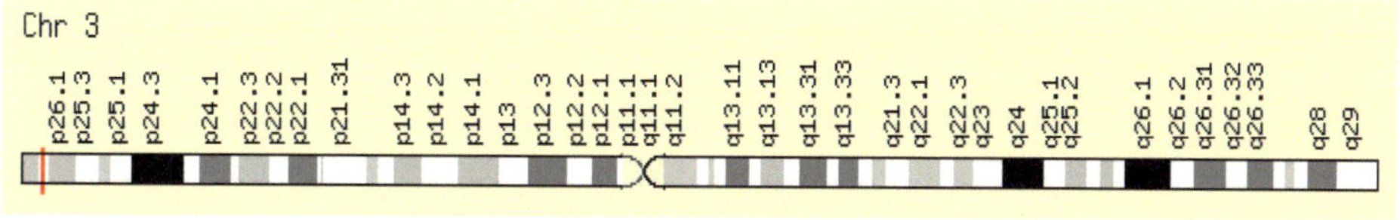

Figure 4.
Molecular location of IL-5 Rα gene. Cytogenetic location, Chr. 3p24-3p26.1.

3. Conclusion

The IL-5 C703T polymorphism could be an important risk factor for asthma in children, while IL-5Rα G80-A genotypes aren't associated with asthma in children. The IL-5 C703T gene polymorphism in the promoter sequences may result in the altered IL-5 production that leads to altered inflammatory responses by increased absolute eosinophil count and, hence, contributes to the pathogenesis of asthma. Serum IL-5 level had a positive linear correlation with eosinophil count. IL-5 levels, absolute eosinophil count and total serum IgE were not associated with IL-5 Rα G-80A polymorphism.

Author details

Raghdah Maytham Hameed[1*], Haidar Abdul Amir Najim Abood[2] and Mohanad Mohsin Ahmed[3]

1 Department of Medical Microbiology, AL-Sader Medical City, Iraq

2 Department of Pharmacology, University of Kerbala, Iraq

3 Department of Medical Microbiology, University of Kerbala, Iraq

*Address all correspondence to: raghdaanalyst92@gmail.com

References

[1] Perez T, Chanez P, Dusser D, Devillier P. Small airway impairment in moderate to severe asthmatics without significant proximal airway obstruction. Respiratory Medicine. 2013;**107**:1667-1674

[2] Fu LS, Tsai MC. Asthma exacerbation in children: A practical review. Pediatrics and Neonatology. 2014;**55**:83-91

[3] Beasley R, Semprini A, Mitchell EA. Risk factors for asthma: Is prevention possible? The Lancet. 2015;**386**:1075-1085

[4] Looijmans-Van Den Akker I, Van Luijn K, Verheij T. Overdiagnosis of asthma in children in primary care: A retrospective analysis. The British Journal of General Practice. 2016;**66**:e152-e157

[5] Papi A, Brightling C, Pedersen SE, Reddel HK. Asthma. Lancet. 2018;**391**: 783-800

[6] Hargreave F, Nair P. The definition and diagnosis of asthma. Clinical & Experimental Allergy. 2009;**39**:1652-1658

[7] Murdoch JR, Lloyd CM. Chronic inflammation and asthma. Mutation Research, Fundamental and Molecular Mechanisms of Mutagenesis. 2010;**690**:24-39

[8] Mims JW. Asthma: Definitions and pathophysiology. International Forum Allergy Rhinology. 2015;**5**(Suppl. 1):S2-S6

[9] Mccracken JL, Veeranki SP, Ameredes BT, Calhoun WJ. Diagnosis and management of asthma in adults: A review. JAMA. 2017;**318**:279-290

[10] Brusselle G, Bracke K. Targeting immune pathways for therapy in asthma and chronic obstructive pulmonary disease. Annals of the American Thoracic Society. 2014;**11**(Suppl. 5):S322-S328

[11] Mahdaviani SA, Rezaei N, Moradi B, Dorkhosh S, Amirzargar AA, Movahedi M. Proinflammatory cytokine gene polymorphisms among Iranian patients with asthma. Journal of Clinical Immunology. 2009;**29**:57-62

[12] Babusikova E, Jurecekova J, Jesenak M, Evinova A. Association of gene polymorphisms in Interleukin 6 in infantile bronchial asthma. Archivos de Bronconeumología. 2017;**53**:381-386

[13] Holgate ST. Innate and adaptive immune responses in asthma. Nature Medicine. 2012;**18**:673-683

[14] Suarez CJ, Parker NJ, Finn PW. Innate immune mechanism in allergic asthma. Current Allergy and Asthma Reports. 2008;**8**:451-459

[15] Wills-Karp M. Complement activation pathways: A bridge between innate and adaptive immune responses in asthma. Proceedings of the American Thoracic Society. 2007;**4**:247-251

[16] Lloyd CM, Hawrylowicz CM. Regulatory T cells in asthma. Immunity. 2009;**31**:438-449

[17] Hansel TT, Johnston SL, Openshaw PJ. Microbes and mucosal immune responses in asthma. Lancet. 2013;**381**:861-873

[18] Buc M, Dzurilla M, Vrlik M, Bucova M. Immunopathogenesis of bronchial asthma. Archivum Immunologiae et Therapiae Experimentalis (Warsz). 2009;**57**:331-344

[19] Bharadwaj AS, Bewtra AK, Agrawal DK. Dendritic cells in allergic

airway inflammation. Canadian Journal of Physiology and Pharmacology. 2007;**85**:686-699

[20] Lambrecht BN, Hammad H. Asthma: The importance of dysregulated barrier immunity. European Journal of Immunology. 2013;**43**:3125-3137

[21] Yoshimoto T. The Hunt for the Source of Primary Interleukin-4: How We Discovered That Natural Killer T Cells and Basophils Determine T Helper Type 2 Cell Differentiation In Vivo. 2018

[22] Georas SN, Guo J, De Fanis U, Casolaro V. T-helper cell type-2 regulation in allergic disease. European Respiratory Journal. 2005;**26**:1119

[23] Lloyd CM, Hessel EM. Functions of T cells in asthma: More than just T(H)2 cells. Nature Reviews. Immunology. 2010;**10**:838-848

[24] Finkelman FD, Hogan SP, Hershey GK, Rothenberg ME, Wills-karp M. Importance of cytokines in murine allergic airway disease and human asthma. Journal of Immunology. 2010;**184**:1663-1674

[25] Cathomson N, Patel M, Smith A. Lebrikizumab in the personalized management of asthma; 2012.

[26] Tomasiak-Lozowska MM, Bodzenta-Lukaszyk A, Tomasiak M, Skiepko R, Zietkowski Z. The role of interleukin 13 and interleukin 5 in asthma. Postępy Higieny i Medycyny Doświadczalnej. 2010;**64**:146-155

[27] Possa SS, Leick EA, Prado CM, Martins MA, Tibério IFLC. Eosinophilic inflammation in allergic asthma. Frontiers in Pharmacology. 2013;**4**:46

[28] Greenfeder S, Umland SP, Cuss FM, Chapman RW, Egan RW. Th2 cytokines and asthma. The role of interleukin-5 in allergic eosinophilic disease. Respiratory Research. 2001;**2**:71

[29] Takatsu K, Kouro T, Nagai Y. Interleukin 5 in the link between the innate and acquired immune response. Advances in Immunology. 2009;**101**:191-236

[30] Takatsu K. Interleukin-5 and IL-5 receptor in health and diseases. Proceedings of the Japan Academy. Series B, Physical and Biological Sciences. 2011;**87**:463-485

[31] Karlen S, De Boer ML, Lipscombe RJ, Lutz W, Mordvinov VA, Sanderson CJ. Biological and molecular characteristics of interleukin-5 and its receptor. International Reviews of Immunology. 1998;**16**:227-247

[32] Kouro T, Takatsu K. IL-5-and eosinophil-mediated inflammation: From discovery to therapy. International Immunology. 2009;**21**:1303-1309

[33] Matthaei KI, Foster P, Young IG. The role of interleukin-5 (IL-5) in vivo: Studies with IL-5 deficient mice. Memórias do Instituto Oswaldo Cruz. 1997;**92**(Suppl 2):63-68

[34] Stirling RG, Van Rensen EL, Barnes PJ, Chung KF. Interleukin-5 induces CD34(+) eosinophil progenitor mobilization and eosinophil CCR3 expression in asthma. American Journal of Respiratory and Critical Care Medicine. 2001;**164**:1403-1409

[35] Takatsu K. Role of interleukin-5 in immune regulation and inflammation. Nihon Rinsho. 2004;**62**:1941-1951

[36] Randall TD, Lund FE, Brewer JW, Aldridge C, Wall R, Corley RB. Interleukin-5 (IL-5) and IL-6 define two molecularly distinct pathways of B-cell differentiation. Molecular and Cellular Biology. 1993;**13**:3929-3936

[37] Foster PS, Hogan SP, Yang M, Mattes J, Young IG, Matthaei KI, et al. Interleukin-5 and eosinophils as therapeutic targets for asthma. Trends in Molecular Medicine. 2002;**8**:162-167

[38] Garcia G, Taille C, Laveneziana P, Bourdin A, Chanez P, Humbert M. Anti-interleukin-5 therapy in severe asthma. European Respiratory Review. 2013;**22**:251-257

[39] Kotsimbos AT, Hamid Q. IL-5 and IL-5 receptor in asthma. Memórias do Instituto Oswaldo Cruz. 1997;**92** (Suppl. 2):75-91

[40] Pereira E, Goldblatt J, Rye P, Sanderson C, Le Souef P. Mutation analysis of interleukin-5 in an asthmatic cohort. Human Mutation. 1998;**11**:51-54

[41] Yamamoto N, Sugiura H, Tanaka K, Uehara M. Heterogeneity of interleukin 5 genetic background in atopic dermatitis patients: Significant difference between those with blood eosinophilia and normal eosinophil levels. Journal of Dermatological Science. 2003;**33**:121-126

[42] Freidin MB, Kobyakova OS, Ogorodova LM, Puzyrev VP. Association of polymorphisms in the human IL4 and IL5 genes with atopic bronchial asthma and severity of the disease. International Journal of Genomics. 2003;**4**:346-350

[43] Boulay J-L, Paul WE. The interleukin-4 family of lymphokines. Current Opinion in Immunology. 1992;**4**:294-298

[44] NCBI. Interleukin 5 [Homo sapiens (human)]. 2019. Available from: https://www.ncbi.nlm.nih.gov/gene/3567#gene-expression. [Accessed: November 16, 2019]

[45] Hameed R, Ahmed M, Abood H. Specific IL-5 snp is associated with high serum IL-5 levels and higher eosinophil counts among iraqi asthmatic children. Biomedical and Biotechnology Research Journal (BBRJ). 2019;**3**:156-161

[46] Losol P, Kim S-H, Seob Shin Y, Min Ye Y, Park H-S. A genetic effect of IL-5 receptor α polymorphism in patients with aspirin-exacerbated respiratory disease. Experimental and Molecular Medicine. 2013;**45**:e14

[47] Hameed R, Najim Abood H, Ahmed M. Systematic review and meta-analysis of the association between C-746T and C-703T polymorphisms of the interleukin-5 gene and asthma. Biomedical and Biotechnology Research Journal (BBRJ). 2020;**4**:323-329

[48] De Vooght KMK, Van Wijk R, Van Solinge WW. Management of gene promoter mutations in molecular diagnostics. Clinical Chemistry. 2009;**55**:698

[49] Tavernier J, Van Der Heyden J, Verhee A, Brusselle G, Van Ostade X, Vandekerckhove J, et al. Interleukin 5 regulates the isoform expression of its own receptor α-subunit. Blood. 2000;**95**:1600-1607

[50] Klion AD, Wilson TM, Brown M, Maric I, Metcalfe DD, Nutman TB. Surface and soluble IL-5 receptor alpha expression in patients with eosinophilia and mastocytosis. Journal of Allergy and Clinical Immunology. 2009;**123**:S251

[51] Cornelis S, Plaetinck G, Devos R, Van der Heyden J, Tavernier J, Sanderson CJ, et al. Detailed analysis of the IL-5-IL-5R alpha interaction: Characterization of crucial residues on the ligand and the receptor. The EMBO Journal. 1995;**14**:3395-3402

[52] Gevaert P, Hellman C, Lundblad L, Lundahl J, Holtappels G, Van

Cauwenberge P, et al. Differential expression of the interleukin 5 receptor alpha isoforms in blood and tissue eosinophils of nasal polyp patients. Allergy. 2009;**64**:725-732

[53] Namkung JH, Lee JE, Kim E, Cho HJ, Kim S, Shin ES, et al. IL-5 and IL-5 receptor alpha polymorphisms are associated with atopic dermatitis in Koreans. Allergy. 2007;**62**:934-942

[54] Wilson TM, Maric I, Shukla J, Brown M, Santos C, Simakova O, et al. IL-5 receptor α levels in patients with marked eosinophilia or mastocytosis. The Journal of Allergy and Clinical Immunology. 2011;**128**:1086-1093

[55] Koike M, Nakamura K, Furuya A, Iida A, Anazawa H, Takatsu K, et al. Establishment of humanized anti-interleukin-5 receptor alpha chain monoclonal antibodies having a potent neutralizing activity. Human Antibodies. 2009;**18**:17-27

[56] Castro M, Wenzel SE, Bleecker ER, Pizzichini E, Kuna P, Busse WW, et al. Benralizumab, an anti-interleukin 5 receptor alpha monoclonal antibody, versus placebo for uncontrolled eosinophilic asthma: A phase 2b randomised dose-ranging study. The Lancet Respiratory Medicine. 2014;**2**:879-890

[57] Tuypens T, Plaetinck G, Baker E, Sutherland G, Brusselle G, Fiers W, et al. Organization and chromosomal localization of the human interleukin 5 receptor alpha-chain gene. European Cytokine Network. 1992;**3**:451-459

[58] NCBI. Interleukin 5 receptor subunit alpha [*Homo sapiens* (human)] [Online]. 2019. Available from: https://www.ncbi.nlm.nih.gov/gene/3568. [Accessed: November 16, 2019]

[59] Cheong HS, Kim LH, Park BL, Choi YH, Park H-S, Hong S-J, et al. Association analysis of interleukin 5 receptor alpha subunit (IL5RA) polymorphisms and asthma. Journal of Human Genetics. 2005;**50**:628

[60] Sehmi R, Wood LJ, Watson R, Foley R, Hamid Q, O'Byrne PM, et al. Allergen-induced increases in IL-5 receptor alpha-subunit expression on bone marrow-derived CD34+ cells from asthmatic subjects. A novel marker of progenitor cell commitment towards eosinophilic differentiation. The Journal of Clinical Investigation. 1997;**100**:2466-2475

[61] Sun Z, Yergeau DA, Wong IC, Tuypens T, Tavernier J, Paul CC, et al. Interleukin-5 receptor alpha subunit gene regulation in human eosinophil development: Identification of a unique cis-element that acts lie an enhancer in regulating activity of the IL-5R alpha promoter. Current Topics in Microbiology and Immunology. 1996;**211**:173-187

[62] Hameed R, Abood H, Ahmed M. Association between the interleukin-5 receptor alpha-subunit G-80A polymorphism and immunological parameters in asthmatic children. Biomedical and Biotechnology Research Journal (BBRJ). 2020;**4**:132-136